THE END OF EVERYTHING

(ASTROPHYSICALLY SPEAKING)

KATIE MACK

SCRIBNER

New York   London   Toronto   Sydney   New Delhi

Scribner
An Imprint of Simon & Schuster, Inc.
1230 Avenue of the Americas
New York, NY 10020

First Scribner hardcover edition August 2020

SCRIBNER and design are registered trademarks of The Gale Group, Inc.,
used under license by Simon & Schuster, Inc., the publisher of this work.

For information about special discounts for bulk purchases,
please contact Simon & Schuster Special Sales at 1-866-506-1949
or business@simonandschuster.com.

The Simon & Schuster Speakers Bureau can bring authors to your live event.
For more information or to book an event, contact the Simon & Schuster Speakers
Bureau at 1-866-248-3049 or visit our website at www.simonspeakers.com.

Diagrams by Nick James

Manufactured in the United States of America

1   3   5   7   9   10   8   6   4   2

Library of Congress Cataloging-in-Publication Data is available.

ISBN 978-1-9821-0354-5
ISBN 978-1-9821-0356-9 (ebook)

For my mother,
who's been there from the beginning

The author is grateful to the Alfred P. Sloan Foundation Public Understanding of Science program for their generous support for the research and writing of this book.

Contents

THE END OF EVERYTHING

CHAPTER 1:

Introduction to the Cosmos

Some say the world will end in fire,
Some say in ice.
From what I've tasted of desire
I hold with those who favor fire.
But if it had to perish twice,
I think I know enough of hate
To say that for destruction ice
Is also great
And would suffice.

Robert Frost, 1920

The question of how the world will end has been the subject of speculation and debate among poets and philosophers throughout history. Of course, now, thanks to science, we know the answer: it's fire. Definitely, fire. In about five billion years, the Sun will swell to its red giant phase, engulf the orbit of Mercury and perhaps Venus, and leave the Earth a charred, lifeless, magma-covered rock. Even this sterile smoldering remnant is likely fated to eventually spiral into the Sun's outer layers and disperse its atoms in the churning atmosphere of the dying star.

So: fire. That's settled. Frost was right the first time.

But he wasn't thinking big enough. I'm a cosmologist. I study the universe, as a whole, on the largest scales. From that perspective, the world is a small sentimental speck of dust lost in a vast and varied universe. What matters to me, professionally and personally, is a bigger question: how will the *universe* end?

We know it had a beginning. About 13.8 billion years ago, the universe went from a state of unimaginable density, to an all-encompassing cosmic fireball, to a cooling, humming fluid of matter and energy, which laid down the seeds for the stars and galaxies we see around us today. Planets formed, galaxies collided, light filled the cosmos. A rocky planet orbiting an ordinary star near the edge of a spiral galaxy developed life, computers, political science, and spindly bipedal mammals who read physics books for fun.

But what's next? What happens at the end of the story? The death of a planet, or even a star, might in principle be survivable. In billions of years, humanity could still conceivably exist, in some perhaps unrecognizable form, venturing out to distant reaches of space, finding new homes and building new civilizations. The death of the universe, though, is final. What does it mean for us, for everything, if it will all eventually come to an end?

WELCOME TO THE END TIMES

Despite the existence of some classic (and highly entertaining) papers in the scientific literature, I first encountered the term "eschatology," the study of the end of everything, by reading about religion.

Eschatology—or more specifically, the end of the world—provides a way for many of the world's religions to contextualize the lessons of theology and to drive home their meaning with overwhelming force. For all the theological differences between Christianity, Judaism, and Islam, they have in common a vision of the End Times that brings about a final restructuring of the world in which good triumphs over evil and those favored by God are rewarded.* Perhaps the promise of a final judgment serves to somehow make up for the unfortunate fact that our imperfect, unfair, arbitrary physical world cannot be relied upon to make existence good and worthwhile for those who live right. In the same way a novel can be redeemed or retroactively ruined by its concluding chapter, many religious philosophies seem to need the world to end, and to end "justly," for it to have had meaning in the first place.

Of course, not all eschatologies are redemptive, and not all religions predict an end time at all. Despite the hype around late December 2012, the Mayan view of the universe was a cyclic one, as it is in Hindu tradition, with no particular "end" designated. The cycles in these traditions aren't mere repetitions, but are imbued with the possibility that things will be better the next time around: all your suffering in this world is bad, but don't worry, a new world is coming, and it will be unscarred, or perhaps improved, by the iniquities of the present. Secular stories of the end, on the other hand, run the gamut from a nihilist view that nothing matters at all (and that nothingness ultimately prevails) to the heady notion of eternal recurrence, where everything that has happened will happen again, in exactly the same way, forever.† In fact, both these seemingly opposing theories are commonly associated with Friedrich Nietzsche, who, after proclaiming the death of any god that might bring order and meaning to the universe, grappled with the implications of living in a cosmos lacking a final redemption arc.

Nietzsche isn't the only one to have contemplated the meaning of existence, of course. Everyone from Aristotle to

* Exactly how those rewards are doled out, and to whom, is not the part they have in common.

† This view is also espoused, though not explored in philosophical detail, in the classic early-2000s TV series *Battlestar Galactica*.

Lao-Tzu to de Beauvoir to Captain Kirk to Buffy the Vampire Slayer has at one point asked, "What does it all mean?" As of this writing, we have yet to reach a consensus.

Whether or not we subscribe to any particular religion or philosophy, it would be hard to deny that knowing our cosmic destiny must have some impact on how we think about our existence, or even how we live our lives. If we want to know whether what we do here ultimately matters, the first thing we ask is: how will it come out in the end? If we find the answer to that question, it leads immediately to the next: what does this mean for us now? Do we still have to take the trash out next Tuesday if the universe is going to die someday?

I've done my own scouring of theological and philosophical texts, and while I learned many fascinating things from my studies, unfortunately the meaning of existence wasn't one of them. I may just not have been cut out for it. The questions and answers that have always drawn me in most strongly are the ones that can be answered with scientific observation, mathematics, and physical evidence. As appealing as it sometimes seemed to have the whole story and meaning of life written down for me once and for all in a book, I knew I would only ever really be able to accept the kind of truth I could rederive mathematically.

LOOKING UP

Over the millennia since humanity's first ponderings of its mortality, the philosophical implications of the question haven't changed, but the tools we have to answer it have. Today, the question of the future and ultimate fate of all reality is a solidly scientific one, with the answer tantalizingly within reach. It hasn't always been so. In Robert Frost's time, debates still raged in astronomy about whether the universe might be in a steady state, existing unchanging forever. It was an appealing idea, that our cosmic home might be a stable, hospitable one: a safe place in which to grow old. The discovery of the Big Bang and the expansion of the universe, however, ruled that out. Our universe is changing, and we've only just begun to develop the theories and observations to understand exactly how. The developments of the last few years, and even months, are finally allowing us to paint a picture of the far future of the cosmos.

I want to share that picture with you. The best measurements we have are only consistent with a handful of final apocalyptic scenarios, some of which may be confirmed or ruled out by observations we're making right now. Exploring these possibilities gives us a glimpse of the workings of science at the cutting edge, and allows us to see humanity in a new context. One which, in my opinion, can bring a kind of joy even in the face of total destruction. We are a species poised between an awareness of our ultimate insignificance and an ability to reach far beyond our mundane lives, into the void, to solve the most fundamental mysteries of the cosmos.

To adapt a line from Tolstoy, every happy universe is the same; every unhappy universe is unhappy in its own way. In this book, I describe how small tweaks to our current, incomplete knowledge of the cosmos can result in vastly different paths into the future, from a universe that collapses on itself, to one that rips itself apart, to one that succumbs by degrees to an inescapable expanding bubble of doom. While we explore the evolution of our modern understanding of the universe and its ultimate end, and grapple with what that means *for us*, we'll encounter some of the most important concepts in physics and see how these connect not just to cosmic apocalypses,* but also to the physics of our everyday lives.

* apocalypsi?

QUANTIFYING COSMIC DOOM

Of course, for some of us, cosmic apocalypses are already a daily concern.

I remember vividly the moment I found out that the universe might end at any second. I was sitting on Professor Phinney's living room floor with the rest of my undergraduate astronomy class for our weekly dessert night, while the professor sat on a chair with his three-year-old daughter on his lap. He explained that the sudden space-stretching expansion of the early universe, cosmic inflation, was still such a mystery that we don't have any idea why it started or why it ended, and we have no way of saying that it won't happen again, right now. No assurance existed to tell us that a rapid, un-survivable rending of space couldn't start right then, in that living room, while we innocently ate our cookies and drank our tea.

I felt completely blindsided, as if I could no longer trust the solidity of the floor beneath me. Forever etched into my brain is the image of that little child sitting there, fidgeting obliviously in a suddenly unstable cosmos, while the professor gave a little smirk and moved on to another topic.

Now that I'm an established scientist, I understand that smirk. It can be morbidly fascinating to ponder processes so powerful and unstoppable yet precisely mathematically describable. The possible futures of our cosmos have been delineated, calculated, and weighted by likelihood based on the best available data. We may not know for certain if a violent new cosmic inflation could occur right now, but if it does, we have the equations ready. In a way, this is a deeply affirming thought: even though we puny helpless humans have no chance of being able to affect (or effect) an end of the cosmos, we can begin to at least understand it.

Many other physicists get a little blasé about the vastness of the cosmos and forces too powerful to comprehend. You can reduce it all to mathematics, tweak some equations, and get on with your day. But the shock and vertigo of the recognition of the fragility of everything, and my own powerlessness in it, has left its mark on me. There's something about taking the opportunity to wade into that cosmic perspective that is both terrifying and hopeful, like holding a newborn infant and feeling the delicate balance of the tenuousness of life and the potential for not-yet-imagined greatness. It is said that astronauts returning from space carry with them a changed perspective on the world, the "overview effect," in which, having seen the Earth from above, they can fully perceive how fragile our little oasis is and how unified we ought to be as a species, as perhaps the only thinking beings in the cosmos.

For me, thinking about the ultimate destruction of the universe is just such an experience. There's an intellectual luxury in being able to ponder the farthest reaches of deep time, and in having the tools to speak about it coherently. When we ask the question, "Can this all really go on forever?," we are implicitly validating our own existence, extending it indefinitely into the future, taking stock, and examining our legacy. Acknowledging an ultimate end gives us context, meaning, even hope, and allows us, paradoxically, to step back from our petty day-to-day concerns and simultaneously live more fully in the moment. Maybe this can be the meaning we seek.

We're definitely getting closer to an answer. Whether or not the world is at any given moment falling apart from a political perspective, scientifically we are living in a golden age. In physics, recent discoveries and new technological and theoretical tools are allowing us to make leaps that were previously impossible. We've been refining our understanding of the beginning of the universe for decades, but the scientific exploration of how the universe might end is just now undergoing its renaissance. Hot-off-the-presses results from powerful telescopes and particle colliders have suggested exciting (if terrifying) new possibilities and changed our perspective on what is likely, or not, in the far future evolution of the cosmos. This is a field in which incredible progress is being made, giving us the opportunity to stand at the very edge of the abyss and peer into the ultimate darkness. Except, you know, quantifiably.

As a discipline within physics, the study of cosmology isn't really about finding meaning per se, but it is about uncovering fundamental truths. By precisely measuring the shape of the universe, the distribution of matter and energy within it, and the forces that govern its evolution, we find hints about the deeper structure of reality. We might tend to associate leaps forward in physics with experiments in laboratories, but much of what we know about the fundamental laws governing the natural world comes not from the experiments themselves, but from understanding their relationship to observations of the heavens. Determining the structure of the atom, for example, required physicists to connect the results of radioactivity experiments with the patterns of spectral lines in the light from the Sun. The Law of Universal Gravitation, developed by Newton, posited that the same force that makes a block slide down an inclined plane keeps the Moon and planets in their orbits. This led, ultimately, to Einstein's General Theory of Relativity, a spectacular reworking of gravity, whose validity was confirmed not by measurements on Earth, but by observations of Mercury's orbital quirks and the apparent positions of stars during a total solar eclipse.

Today, we are finding that the particle physics models we've developed through decades of rigorous testing in the best Earthly laboratories are incomplete, and we're getting these clues from the sky. Studying the motions and distributions of other galaxies—cosmic conglomerations like our own Milky Way that contain billions or trillions of stars—has pointed us to major gaps in our theories of particle physics. We don't know yet what the solution will be, but it's a safe bet that our explorations of the cosmos will play a role in sorting it out. Uniting cosmology and particle physics has already allowed us to measure the basic shape of spacetime, take an inventory of the components of reality, and peer back through time to an era before the existence of stars and galaxies in order to trace our origins, not just as living beings, but as matter itself.

Of course, it goes both ways. As much as modern cosmology informs our understanding of the very, very small, particle theories and experiments can give us insight into the workings of the universe on the largest scales. This combination of a top-down and bottom-up approach ties into the essence of physics. As much as pop culture would have you believe that science is all about eureka moments and spectacular conceptual reversals, advances in our understanding come more often from taking existing theories, pushing them to the extremes, and watching where they break. When Newton was rolling balls down hills or watching the planets inch across the sky, he couldn't possibly have guessed that we'd need a theory of gravity that could also cope with the warping of spacetime near the Sun, or the unimaginable gravitational forces inside black holes. He would never have dreamed that we'd someday hope to measure the effect of gravity on a single neutron.* Fortunately, the universe, being really very big, gives us a lot of extreme environments to observe. Even better, it gives us the ability to study the early universe, a time when the entire cosmos was an extreme environment.

• • •

---

* We do this by bouncing it. Really. First we cool the neutrons to almost absolute zero, then we slow them to jogging speed, then we bounce them up and down like a Ping-Pong ball on a paddle. And this also tells us something about dark energy, the mysterious something that makes our whole universe expand faster. Physics is wild.

A quick note about terminology. As a general scientific term, *cosmology* refers to the study of the universe as a whole, from beginning to end, including its components, its evolution over time, and the fundamental physics governing it. In *astrophysics*, a cosmologist is anyone who studies really distant things, because (1) that means looking at quite a lot of universe and (2) in astronomy, faraway things are also far in the past, since the light that reaches us from them has been traveling for a long time—sometimes billions of years. Some astrophysicists explicitly study the evolution or early history of the universe, while some specialize in distant objects (galaxies, clusters of galaxies, and so forth) and their properties. In *physics*, cosmology can veer in a direction that is much more theoretical. For instance, some cosmologists in physics departments (as opposed to astronomy departments) study alternative formulations of particle physics that might have applied to the first billionth of a billionth of a second of the universe's existence. Others study modifications of Einstein's theory of gravity that could pertain to objects as hypothetical as black holes that can only exist in higher dimensions of space. Some cosmologists even study whole hypothetical universes that are very explicitly not our own—universes in which the cosmos has a totally different shape, number of dimensions, and history—in order to gain insight into the mathematical structure of theories that *might* someday be found to have relevance to us.*

The upshot of all this is that cosmology means a lot of dif-

---

* String theorists produce a lot of these theories. (String theory is a blanket term for theories that try to bring together gravity and particle physics in new ways, but most of the work done to develop it now relies on mathematical analogs rather than anything pertaining to the "real" world.) Sometimes when I'm in string theory talks, I have to resist the urge to raise my hand and clarify that none of these calculations pertain to *our* universe, just in case anyone in the room is as confused as I first was when I started attending string theory talks.

ferent things to a lot of different people. A cosmologist who studies the evolution of galaxies might be utterly lost talking with a cosmologist who studies the way quantum field theory can make black holes evaporate, and vice versa.

As for me, I love it all. I first learned cosmology was a thing when I was about ten years old, through encounters with books and lectures by Stephen Hawking. He was talking about black holes and warped spacetime and the Big Bang and all sorts of stuff that made me feel like my brain was doing backflips. I *could not get enough*. When I found out that Hawking described himself as a cosmologist, I knew that was what I wanted to be. Through the years, I've done research across the whole range, bouncing back and forth between physics and astronomy departments, studying black holes, galaxies, intergalactic gas, intricacies of the Big Bang, dark matter, and the possibility that the universe might suddenly blink out of existence.* I even dabbled in experimental particle physics for a while, in my misspent youth, playing with lasers in a nuclear physics lab (despite what the records might say, the fire was not my fault) and paddling an inflatable boat around a 40-meter-tall water-filled underground neutrino detector (that explosion was not my fault either).

These days, I'm pretty solidly a theorist, which is probably better for everyone. This means I don't carry out observations or experiments or analyze data, though I do frequently make predictions for what future observations or experiments might see. I work mainly in an area physicists call phenomenology—the space between the development of new theories and the part where they're actually tested. That is to say, I find creative new ways to connect the things the fundamental-theory people hypothesize about the structure of the universe with

---

* This is, of course, one of the most fun things I've ever worked on, hence this book. I'm not sure why I like it so much. It may be a bad sign.

what the observational astronomers and experimental physicists hope to see in their data. It means I have to learn a lot about everything,* and it's a heck of a lot of fun.

SPOILER ALERT

This book is an excuse for me to dig deep into the question of where it's all going, what that all means, and what we can learn about the universe we live in by asking these questions. There isn't just one accepted answer to any of this—the question of the fate of all existence is still an open one, and an area of active research in which the conclusions we draw can change drastically in response to very small tweaks in our interpretations of the data. In this book, we'll explore five possibilities, chosen based on their prominence in ongoing discussions among professional cosmologists, and dig into the best current evidence for or against each of them.

Each scenario presents a very different style of apocalypse, with a different physical process governing it, but they all agree on one thing: there will be an end. In all my readings, I have not yet found a serious suggestion in the current cosmological literature that the universe could persist, unchanged, forever. At the very least, there will be a transition that for all intents and purposes destroys *everything*, rendering at least the observable parts of the cosmos uninhabitable to any organized structure. For this purpose, I will call that an ending (with apologies to any temporarily sentient bursts of random quantum fluctuation† that may be reading this). A few of the scenarios carry with them a hint of possibility that the cosmos might renew

---

* And we're talking about the universe here, so I really do mean EVERYTHING.

† Please stick around until Chapter 4, when the Boltzmann Brain community will get their proper due.

itself, or even repeat, in one way or another, but whether some tenuous memory of previous iterations can persist in any way is a matter of rather intense ongoing debate, as is whether or not anything like an escape from a cosmic apocalypse could in principle be possible. What seems most likely is that the end for our little island of existence known as the observable universe is, truly, the end. I'm here to tell you, among other things, how that might happen.

Just to get everyone on the same page, we'll start with a quick catch-up on the universe from the beginning until now. Then we'll get on with the destruction. In each of five chapters, we'll explore a different possibility for the end, how it might come about, what it would look like, and how our changing knowledge of the physics of reality leads us from one hypothesis to another. We'll start with the Big Crunch, the spectacular collapse of the universe that would occur if our current cosmic expansion were to reverse course. Then come two chapters of dark-energy-driven apocalypses, one in which the universe expands forever, slowly emptying and darkening, and one in which the universe literally rips itself apart. Next is vacuum decay, the spontaneous production of a *quantum bubble of death** that devours the cosmos. Finally, we'll venture into the speculative territory of cyclic cosmology, including theories with extra dimensions of space, in which our cosmos might be obliterated by a collision with a parallel universe . . . over and over again. The closing chapter will bring it all together with an update from several experts currently working on the cutting edge on which scenario looks most plausible now, and what we can expect to learn from new telescopes and experiments to settle the question once and for all.

What that means for us as human beings, living our lit-

---

* Technically it is called a "bubble of true vacuum," which, to be fair, also sounds pretty darn ominous.

tle lives in all this inconsiderate vastness, is another question entirely. We'll present a range of perspectives in the epilogue, and address whether or not sentience itself could have any kind of legacy that endures beyond our destruction.*

We don't know yet whether the universe will end in fire, ice, or something altogether more outlandish. What we do know is that it's an immense, beautiful, truly awesome place, and it's well worth our time to go out of our way to explore it. While we still can.

*Another spoiler: it's not looking great.

CHAPTER 2:

Big Bang to Now

Beginnings imply and require endings.
Ann Leckie, *Ancillary Justice*

I love stories about time travel. It's easy to quibble about the physics of time machines or to balk at the various paradoxes that come up. But there's something appealing about the idea that we might somehow find a trick that will open up the past and future to our knowledge and intervention, to allow us to step off this runaway train of "now" barreling inexorably toward some unknown fate. Linear time just seems so restrictive, even wasteful—why should all that time, all those possibilities, be lost to us forever, just because the clock has ticked forward a few degrees? We may have grown accustomed to strict chronological oppression, but that doesn't mean we have to like it.

Fortunately, cosmology can help. Not in any practical sense, of course—we're still talking about a relatively esoteric branch of physics that will in no way enable you to get back the umbrella you left on the train yesterday. But rather, in the sense that your life remains the same but absolutely everything else about existence is forever changed.

To a cosmologist, the past is not some unreachable lost realm. It's an actual place, an observable region of the cosmos, and it's where we spend most of our workday. We can, while sitting quietly at our desks, watch the progress of astronomical events that happened millions or even billions of years ago. And the trick isn't special to cosmology, but inherent to the structure of the universe in which we live.

It all comes down to the fact that light takes time to travel. Light speed is fast—about 300 million meters per second—but it's not instantaneous. In everyday terms, when you switch on a flashlight, the light coming out of it covers about one foot per nanosecond, and the reflection of that light off whatever you're illuminating takes just as long to get back to you. In fact, when you look at anything, the image you see, which is just the light coming off it that reaches your eye, is a little bit stale by the time it gets to you. That person sitting across the café from you is, from your perspective, several nanoseconds in the past, which may go part of the way toward explaining their wistful expression and outdated fashion sense. Everything you see is in the past, as far as you're concerned. If you look up at the Moon, you're seeing a little over a second ago. The Sun is more than eight minutes in the past. And the stars you see in the night sky are deep in the past, from just a few years to millennia.

The concept of this kind of light speed delay might already be familiar to you, but its implications are profound. It means that as astronomers, we can look into the sky and watch the evolution of the universe happen, from its early beginnings to the present day. We use the unit "light-year" in astronomy not just because it's conveniently huge (about 9.5 trillion kilometers, or 5.9 trillion miles), but also because it tells us how long light has been traveling from the thing we're looking at. A star 10 light-years away is 10 years in the past, from our perspective. A galaxy 10 billion light-years away is 10 billion years in the past. Since the universe is only about 13.8 billion years old, that 10-billion-light-year-distant galaxy can tell us about the conditions of our universe when it was still in its youth. In that sense, looking out into the cosmos is tantamount to looking into our own past.

There's an important caveat to this, and I would be remiss if I didn't mention it. We technically can't see our *own* past at all. The light speed delay means that the more distant a thing is, the farther in the past it is, and that relationship is strict: not only can we not see our own past, we can't see those distant galaxies in the present, either. The more distant something is, the farther away it is on a timeline of the cosmos.

So how do we learn anything useful about our own past if we're only seeing the past for some other galaxy, long ago and far away? It comes down to a principle so central to cosmology that it is literally called the *cosmological principle*. Simply stated, it's the idea that for all practical purposes, the universe is basically the same everywhere. Obviously this isn't true on human scales—the surface of the Earth is pretty importantly different from deep space or the center of the Sun—but on the kind of astronomically large scales on which whole galaxies can be tallied up as individually uninteresting specks, the universe looks the same in every direction, and is made of all the

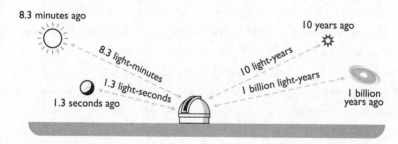

Figure 1: Light travel times. We sometimes express distances in light-seconds, light-minutes, and light-years because it makes it clear how long the light has been traveling to us, and thus how far into the past we're looking. (None of the illustrations here are to scale!)

same stuff.* This idea is closely related to the Copernican Principle, which is the erstwhile heretical notion stated by Nicolaus Copernicus in the sixteenth century that we do not occupy a "special place" in the cosmos, but are just at some generic spot that may as well have been chosen at random. So when we look at a galaxy a billion light-years away, and see it as it was a billion years ago, in a universe that was a billion years younger than our universe is here and now, we can be pretty confident that the conditions *here* a billion years ago would have been fairly similar. This can actually be observationally tested, to a degree. Studies of the distribution of galaxies throughout the cosmos find that the uniformity implied by the cosmological principle holds up everywhere we've looked.

The upshot of all this is that if we want to learn about the evolution of the universe itself, and the conditions our own Milky Way galaxy grew up in, all we have to do is *look at something far away.*

It also means that cosmology doesn't really have a well-defined concept of "now." Or rather, the "now" you experience is highly specific to you, to where you are and to what you are doing.† What does it mean to say "that supernova is going off now" if we see the light of it now, and we can watch the star explode now, but that light has been traveling for millions of years? The thing we're watching is essentially fully in the past, but the "now" for that exploded star is unobservable

---

* Science fiction loves to ignore this. There's an early episode of *Star Trek: The Next Generation* in which they accidentally travel a billion light-years in a few seconds and the place they end up in is some kind of abyss of shimmering blue energy and thought, which, if it really existed, we'd totally be able to see in telescopes.

† We can thank relativity for this. Special relativity says time passes more slowly for us when we are moving quickly; general relativity says it slows down when we are close to a massive object.

to us, and we won't receive any knowledge of it for millions of years, which makes it, to us, not "now," but the future.

When we think of the universe as existing in *spacetime*—a kind of all-encompassing universal grid in which space is three axes and time is a fourth—we can just think of the past and the future as distant points on the same fabric, stretching across the cosmos from its infancy to its end. To someone sitting at a different point on this fabric, an event that is part of the future to us might be the distant past to them. And the light (or any information) from an event that we won't see for millennia is already streaming across spacetime toward us "now." Is that event in the future, or the past, or, perhaps, both? It all depends on perspective.

As mind-bending as it is to contemplate if you're used to thinking in a 3D world,* to astronomers, the noninfinite speed of light is a fantastically useful tool. It means that instead of looking for mere clues to the distant past of the cosmos—its traces and remnants—we can just look at it directly and watch it change over time. We can peer at the universe at an age of just three billion, during the renaissance of star formation, when galaxies were bursting with light (if not art and philosophy), and we can see how that shine has dimmed in the intervening eons. We can look even farther back, and see matter swirling into supermassive black holes in a universe less than 500 million years old, when starlight had only just begun to penetrate the darkness between galaxies.

Soon, with new space telescopes, we will be able to observe some of the first galaxies to form in the cosmos—those that formed when the universe was only a few hundred million years old. But if those galaxies were the first, what happens if

---

* When Doc Brown in *Back to the Future* proclaimed, "You're not thinking fourth-dimensionally!," he was talking to you.

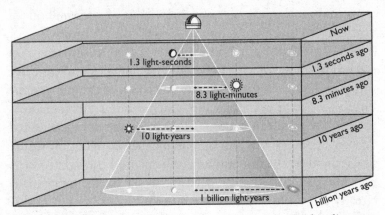

Figure 2: Diagram of light moving through spacetime. In this diagram, time moves forward in the upward direction, and we're showing only two dimensions of space instead of all three. The positions of four objects that are stationary in space are represented by the vertical dotted lines, marking the same location at different times. The "light cone" is the region we can see in the past from the observatory—it encompasses everything close enough to us that the light has had time to reach us since it was emitted. We can see a galaxy a billion light-years away as it was a billion years ago, but we can't see what it looks like "now," because the "now" version of that galaxy is outside our light cone.

we look farther back than that? Can we look so far away that there are no galaxies yet? We have plans to do so. Radio telescopes being built now may be able to see the material the first galaxies were born from, by exploiting a fortuitous interaction between light and hydrogen. By looking directly at the hydrogen, the matter that will one day become stars and galaxies, we can watch the very first structures in the universe form.

But what if we look back even farther? What if we look back to the time before stars, before galaxies, before hydrogen? Can we see the Big Bang itself?

Yes. We can.

SEEING THE BIG BANG

There's a popular picture of the Big Bang as some kind of explosion—a sudden conflagration of light and matter from a single point that billowed out through the universe. It wasn't like that. The Big Bang wasn't an explosion within the universe, it was an expansion *of* the universe. And it didn't happen at a single point, but at *every* point. Every point in space in the universe today—a spot on the edge of a distant galaxy, a piece of intergalactic space just as far in the other direction, the room in which you were born—every one of these points was, at the beginning of time, close enough to touch, and at that same first moment, rapidly tearing away from one another.

The logic of the Big Bang theory is pretty simple. The universe is expanding—we can see that distances between galaxies are getting larger over time—which means that the distances between galaxies were smaller in the past. We can, as a thought experiment, rewind the expansion we see now, extrapolating back across billions of years, until we reach a moment when the distance between galaxies must have been zero. The observable universe, encompassing everything we can see today, must have been contained within a much smaller, denser, hotter space. But the observable universe is just the part of the cosmos we can see now. We know that space goes on much farther than that. In fact, based on what we know, it's entirely possible, and perhaps probable, that the universe is infinite in size. Which means that it was infinite at the beginning too. Just much denser.

This is not easy to picture. Infinities are tough that way. What does it mean to have infinite space? What does it mean for an infinite space to be expanding? How does infinite space get infinit*er*?

I'm afraid I can't help you with this.

There is simply no easy way to hold infinite space in a finite brain. What I can say is that there are ways to deal with infinities in mathematics and physics that make sense and don't break anything. As a cosmologist, I work from the basic assumption that the universe can be described with math, and if that math works out, and is useful for approaching new problems, I go with it.* Or, more precisely, if the math works out and a somewhat different assumption (e.g., that the universe is not *quite* infinite but is so big that we can't possibly ever perceive its edges) also works but makes no difference to our experience or anything we can measure in any way, we may as well stick with the simpler assumption for now. So: infinite universe. We can work with that.

In any case, when we talk about the Big Bang theory, what we're really saying is: based on our observations of the present expansion and its history, we can conclude that there was a time when the universe was, everywhere, much hotter and denser than it is today.† This is sometimes called the "Hot Big Bang," referring to the whole span of time when the universe was hot and dense, which we now know to be the time from year 0 to somewhere around year 380,000.‡

---

* I'm being a bit flippant here but this is a rather important point. So far, in physics, most of what we've done is describe the universe with mathematical constructions we call *models*, and use experiments and observations to test and refine those models, until we arrive at a model that seems to fit the observations better than any of the others. And then we start trying to break the model. It's not that we just trust that math is fundamental to the universe, it's that there doesn't seem to be any other way to approach these things that makes any sense.

† "Our WHOLE universe was in a hot dense state, then nearly 14 billion years ago expansion started . . ." Yes, the Barenaked Ladies got it right: the beginning of the theme song for the TV show *The Big Bang Theory* is actually a very good summary of the theory itself.

‡ Of course, this is before "years" were a thing, because it is before there was a planet orbiting a star, defining a unit of time. But we can take our own units and extrapolate back and just label all the seconds as they count up to years and give them numbers, for our own convenience.

We can even quantify what "hot and dense" means, and trace the history of the universe backward from the cool and pleasant cosmos we are enjoying now to a pressure-cooker inferno so extreme it shatters our understanding of the laws of physics.

This isn't just a theoretical exercise, though. It's one thing to mathematically extrapolate expansion and derive higher pressures and temperatures; it's another to see this infernoverse* directly.

THE COSMIC MICROWAVE BACKGROUND

The story of how we went from thinking about the Big Bang to seeing it is a classic tale of serendipitous discovery in cosmology. In 1965, a physicist named Jim Peebles at Princeton University was doing the calculations, dialing back the cosmic expansion, and coming to the startling conclusion that radiation from the Big Bang should still be streaming through the universe today. Moreover, it should be detectable. He calculated the expected frequency and intensity of that radiation and teamed up with colleagues Robert Dicke and David Wilkinson to start building an instrument to measure it. Meanwhile, unbeknownst to them, just down the road at Bell Labs, a couple of astronomers named Arno Penzias and Robert Wilson were gearing up to do some astronomy with a microwave detector that had previously been used for commercial purposes. (Microwaves are just a kind of light on the electromagnetic spectrum, higher frequency than radio but lower frequency than infrared or visible light.) When Penzias and Wilson, totally uninterested in commercial applications and keen on studying the sky, were calibrating the instrument for their research, they found a weird humming in the feed. Apparently it hadn't interfered with the

---

* I just now invented this term and I'm feeling very proud of myself.

previous use of the telescope, detecting communication signals bounced off high-atmosphere balloons, so those users had ignored it. But this was for *science*, and it had to be fixed. The hum appeared no matter which direction they pointed the detector and was, by all accounts, extremely inconvenient.

Telescope interference is a common problem during the calibration phase of an observing run, and there are a lot of ways it can happen. There could be a loose cable somewhere, or radio interference coming from some transmitter nearby, or any number of little mechanical annoyances. (A recent big break in radio astronomy involved the discovery that tantalizing bursts of radiation seen by the Parkes radio telescope were actually due to an overeager microwave oven in the lunch room.) Penzias and Wilson examined every inch of the detector, and even considered the possibility that a small group of pigeons nesting in the antenna could be the source of the hum.* But no matter what they did, they couldn't get the hum to go away, and they never found any interference that could account for it. So they had to consider the possibility that it really was coming from space, and from every direction in the sky. But what could it be? Anything coming from the planets or the Sun should only show up at certain times and directions, and even emission from our own Milky Way galaxy would not be completely uniform.

Enter the Princeton team. In a roundabout way.

To backtrack a moment, Peebles's calculations had said that if the universe was hot everywhere early on, then we should be currently awash in the leftover radiation from it. Here's what he was thinking. If looking farther away means looking farther into the past, and if there was a time in the distant past when the universe was basically one big all-encompassing fireball, then it should be possible to look so far away that you see

---

* Sadly, this line of investigation did not end well for the pigeons, who were in fact innocent of all wrongdoing.

a part of the universe that is *still on fire*. Or, thinking about it another way: if, 13.8 billion years ago, the whole possibly infinite universe was aglow with radiation, there should be parts of it so far away that the radiation from that glow is only just now reaching us, having traveled through the expanding, cooling space all this time. In any direction we look, if we look far enough away, we'll see that distant fiery universe. We're not looking at parts of *space* that are any different, but rather at a *time* when ALL of space was on fire.

So, this background radiation should come from everywhere. And it should come from everywhere no matter where you are, because you can always look far enough away to see the hot phase of the cosmos. The speed-of-light/time-travel connection gives you this for free. Every point in space is the center of its own sphere of ever deepening time, bounded by a shell of fire.

Peebles realized this, and, as physicists tend to do, talked to his colleagues about his extremely mind-blowing thoughts. He even passed around a preprint of a paper in which he described what he and his colleagues planned to do to detect this radiation. Then, word proceeded to travel the 60 kilometers to Bell Labs—via two unconnected physicists, an airplane, and Puerto Rico.

An attendee of Peebles's talk, Ken Turner, went to visit the Arecibo radio telescope, and on the flight back, had a chat with fellow astronomer Bernard Burke about how cool it would be to detect this Big Bang radiation. After arriving back at the office, Burke got a phone call from Penzias about some unrelated work, and happened to mention the airplane chatter.* At

---

* Without knowing anything about this story except the bit about the pigeons, I happened to run into Bernard Burke a few years ago at MIT. We were just chatting as physicists tend to do, and he was telling me about some past work I didn't really follow and at some point I realized he was talking about his phone call with THE Penzias and just casually dropping the fact that he was a catalyst for one of the most important discoveries in the history

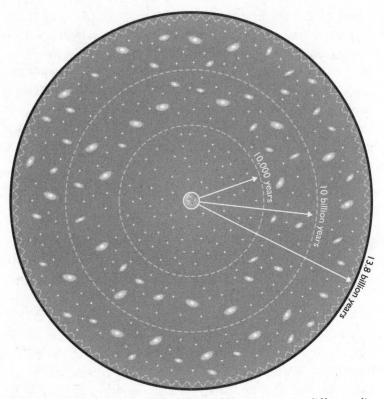

Figure 3: A cartoon map of the observable universe. At different distances from our vantage point on Earth, we can see different epochs in the past. The lookback time (the number of years before today) is labeled for each sphere around us in this diagram. The farthest we can see, even in principle, corresponds to a distance from Earth from which light leaving that point at the very beginning of the universe would be reaching us right now. This defines a sphere around us known as the *observable universe.*

of physics. A similar thing happened to me at a conference several years ago when I met Tom Kibble, who built much of the theory around the Higgs boson. Moral of the story: listen to the emeritus profs; they might have quietly revolutionized your whole field of study.

which point, I can only assume Penzias had to have a bit of a sit-down, because now he knew that he and Wilson had just become the first human beings to *see the actual Big Bang*. He took a couple days, talked with his colleague, and then phoned up Robert Dicke, who immediately turned around to Peebles and Wilkinson and said, "We've been scooped."

And indeed they had. Penzias and Wilson went on to win the Nobel Prize in 1978 for the first observation of what became known as the *cosmic microwave background.**

The cosmic microwave background, or CMB, went on to become one of the most important tools we have for studying the history of the universe. It's hard to overstate its significance, both as an astronomical data set and as a technological achievement. We can now collect, analyze, and map the glow of the hot early cosmos. The first thing it tells us is: the hypothesis that the early universe was one big inferno, glowing with heat, is completely confirmed.

But how do we know for sure that the background light we detect is actually from the primordial fireball, and not from, say, some collection of weird faraway stars or something? It turns out the light's spectrum—the way it's brighter or dimmer when measured at different frequencies—contains a dead giveaway.

Let's say you have a fireplace, and you stick a poker in the fire until it starts to glow red. That red glow isn't a property of the metal itself, it's a phenomenon that happens to anything that gets heated up (without bursting into flames). That glow is called "thermal radiation," and the color of it depends only on the temperature. Something glowing blue is hotter than something glowing red. In fact, if you could see infrared light, you'd see thermal

---

* In the course of writing this book, I was thrilled to hear that Peebles won a 2019 Nobel Prize for, in part, the theory side of this discovery. So maybe there is some justice in the end. Just not for the pigeons.

radiation coming from people and warm food and sun-soaked sidewalks all the time. Human thermal radiation comes out at the low frequency of infrared light because we're much cooler than open flames, unless things are going very badly for us.

The color you see isn't the entirety of the light produced, though. Aside from lasers, anything that produces light makes it at a spread of different frequencies (or colors), and the color it appears to the eye is just the color where the light is most intense. (This is why incandescent light bulbs are hot to the touch: even though much of the light they produce is visible, a lot of extra light is produced in the infrared part of the spectrum, which makes them feel hot.) For any thermal radiation, including that emitted by pokers and people and the little blue flames on gas stovetops, the intensity of the light changes with frequency in exactly the same way. The light is brightest at some peak color depending on the temperature, and it dims quickly for colors on either side. Plotting out on a graph how the intensity rises and falls with frequency gives you a shape we call the *blackbody curve*—a curve reproduced by anything that glows because it is hot.* And it turns out that if you measure the intensity of the cosmic microwave background at different frequencies, you get the most precise, most perfect blackbody curve ever measured in nature. The only way to explain this is that the universe itself was once, everywhere, extremely hot.

Legend has it that the first time this result was presented as a graph in a conference talk, the audience actually cheered. Part of their enthusiasm was certainly because the measurement was extremely impressive and precise, and a perfect fit

---

* The name "blackbody" comes from the idea of an object—a "body"—that perfectly absorbs all the light that hits it and reemits it as pure heat. Most objects don't do this perfectly, of course; they reflect a bit of the light, and some is absorbed without being reemitted. But most materials when heated up will glow at some level in a way that is recognizable as an approximate shape of a blackbody curve.

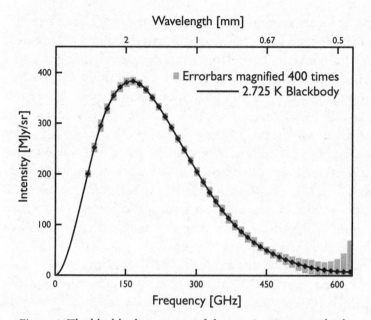

Figure 4: The blackbody spectrum of the cosmic microwave background. The height of the curve indicates the intensity of the radiation at a given frequency or wavelength. The data points are shown with error bars indicating the uncertainties in the measurement, but with the size of the uncertainties magnified 400 times so they're not all completely hidden behind the width of the line, which is the spectrum you'd expect for something glowing at a temperature of 2.725 Kelvins (-270°C).

to the theory (which is always nice to see). But I'm fairly sure that another part was the fact that people realized they were SEEING THE BIG BANG. Actually *seeing* it. I, for one, am still not quite over this.

Aside from the mind-blowing-ness of it, the CMB gives us an invaluable window into the universe's first moments, and into how it has grown and evolved over time. It also gives us some hints about where it is all going, as we'll see in later chapters.

That said, when you make a map of the CMB that shows

the color variation of the light across the sky, it actually looks fairly boring: it is *very nearly* exactly the same color everywhere. The minute deviations you can detect, though, as tiny as they are, tell us a lot. When they crank up the contrast enough to get some color variation, astronomers can see that the CMB looks ever so slightly blotchy, as if someone did an abstract pointillism painting on the sky with a brush as big around as the full Moon viewed from Earth. These blotches collect in clumps of one color in some places and mix in others, with some of the spots a little tiny bit redder, some a little tiny bit bluer.* The color variations reveal places where the roiling primordial cosmic plasma was just slightly cooler or hotter, due to very, very slight changes in density—each point has a density that deviates from the average by no more than about one part in 100,000. (For some perspective on what one part in 100,000 looks like, empty a soda can into a backyard swimming pool.)

We can, with careful calculations, work out how those tiny variations in density are destined to grow over time, starting from minuscule blips and, over the millennia, growing up into entire clusters of galaxies. Gravitational collapse is a powerful thing. If you have a little bit of matter that's denser than the matter around it, it will pull in more from those less dense places, which increases the contrast, which pulls in more, and so on. The rich get richer and the poor get poorer.

Using computer simulations that show billions of years passing in the span of seconds, we can watch as a patch of matter that is only a tiny fraction denser than the one next to it pulls in enough of its surrounding gas to form the first star in the universe. These stars form within the first galaxies,

---

* The light is all in the microwave part of the spectrum, so "redder" means lower-frequency microwave radiation, and "bluer" means higher-frequency microwave radiation, but when we make maps we do actually use colors like red and blue, because, you know, human eyes.

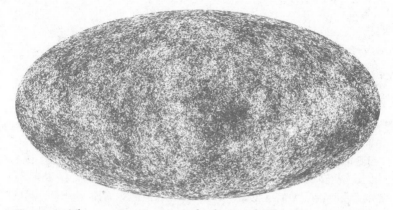

Figure 5: The cosmic microwave background. This is a microwave-frequency map of the whole sky projected onto an oval in a Mollweide projection (with the emission from our own galaxy removed). The darker regions indicate slightly cooler (lower frequency, or redder) microwave emission and the light regions are slightly hotter (higher frequency, or bluer) emission. These indicate, respectively, parts of the early universe that were slightly denser or slightly less dense than their surroundings, at a level of one part in 100,000.

which gather into clusters of galaxies, which en masse turn the splotchy patchwork of the CMB into what we now see as a cosmic web: a veiny arrangement of nodes and filaments and voids traced out by the galaxies shining along it, like dew drops on a spider's web. If you compare the results of one of these simulations with an actual map of the cosmos, where each galaxy is one point in a giant 3D map, they are in such incredible agreement that you won't be able to tell the difference.

So. The Big Bang happened. We've seen it, we've calculated it, the physics adds up. Now, let's all gather around in the glow of the cosmic blackbody and tell the origin story of the cosmos.

IN THE BEGINNING

Not all of cosmic history is as directly visible as the cosmic microwave background. The few hundred thousand years before the end of the fireball stage, and the half million or so years right after, are extremely hard to observe. In the former case, it's because of too much light (imagine trying to look through a wall of fire) and in the latter, because of too little (imagine trying to look at some specks of dust in the air between you and a wall of fire). But the CMB, right in the middle, gives us a solid anchor to extrapolate from in each direction, and now we have a compelling narrative for how the universe evolved over time, starting from the first billionth of a billionth of a billionth of a second and arriving at today, 13.8 billion years later.

Shall we?

In the beginning, there was the singularity.

Well, maybe. A singularity is what most people think of when they think of the Big Bang: an infinitely dense point from which everything in the universe exploded outward. Only, a singularity doesn't have to be a point—it could just be an infinitely dense state of an infinitely large universe. And, as discussed above, there's no explosion per se, since an explosion implies an expansion *into something*, rather than an expansion *of everything*. The idea that everything started with a singularity comes from observing the current expansion of the universe, applying Einstein's equations of gravity, and extrapolating backward. But that singularity might never have happened. What most physicists do think happened, a fraction of a second after whatever was the true "beginning," was a dramatic super-expansion that effectively erased all trace of whatever went on before it. So the singularity is one hypothesis for what might have started everything off, but we can't really be sure.

There's also the question of what was "before" the singularity. That question might be, depending on who you ask, incoherent nonsense (because the singularity was the beginning of time as well as space, so it had no "before"), or one of the most crucial questions in cosmology (because the singularity might have been the end point of a previous phase of a cyclic universe: one that goes from Big Bang to Big Crunch and back again forever). We'll talk about the latter possibility in Chapter 7, but in the meantime, there's not much to say about the singularity other than that it might have happened. Even if we did trust ourselves to dial back expansion all the way to that point, a singularity represents a state of matter and energy so extreme that nothing we currently know about physics can describe it.

To a physicist, a singularity is pathological. It's a place in the equations where some quantity that is normally well behaved (like the density of matter) goes to infinity, at which point there is no longer any way to calculate things that makes any sense. Most of the time, when you encounter a singularity, it is telling you that something has gone wrong in your calculations and you need to go back to the drawing board. Finding a singularity in your theory is like having your satellite navigation direct you to the edge of a lake, then instruct you to disassemble your car, reassemble it as a boat, and paddle your new car-boat to the other side. Maybe this really is the only way to get where you're trying to go, but more likely, you've made a wrong turn somewhere a few miles back.

In practice, though, it doesn't even take something as clearly dysfunctional as a true singularity to lay waste to physics as we know it. Any time you have a lot of energy in a very small space, you have to deal with both quantum mechanics (the theory governing particle physics) and general relativity (the theory of gravity) at the same time. Under normal circumstances, you're only dealing with one or the other, because when gravity is important, it's usually because you have a massive thing, so you can ignore the individual particles, and when quantum mechanics is important, at the particle scale, you're dealing with so little mass that gravity is a totally negligible part of the interaction. But at extreme densities you have to contend with both, and they don't work well together, *at all*. Extreme gravity involves well-defined massive objects that warp space and alter the flow of time; quantum mechanics allows particles to pass through solid walls or exist only as fuzzy probability clouds. The fundamental incompatibility of our theories of the very massive and the very small is one of the things that hints at the direction we should go in creating new, more complete theories. But it is also rather inconvenient when we're trying to explain the very early universe.

Without a full theory of quantum gravity (something that reconciles particle physics with gravity), there's a limit to how far back we can extrapolate the universe in a way that makes sense. We inevitably reach a moment before which all bets are off. During that time, the densities are high enough that we expect extreme gravitational effects to be competing with the inherent fuzziness of quantum mechanics, and we just don't know what to do in that scenario. Do microscopic black holes form (because of the strong gravity) but then pop in and out of existence at random (because of the quantum uncertainty)? Does time have any meaning when the shape of space is no more predictable than a roll of dice? If you zoom in to a small enough scale, do space and time act like discrete particles, or perhaps waves that interfere with each other? Are there wormholes? Are there dragons??? We have no idea.

But because we need to quantify exactly how confused we are, and at what moment that confusion sets in, we call this the Planck Time,* and it encompasses the time from zero to about 10^{-43} seconds. If you're not familiar with the notation, 10^{-43} seconds is equal to one second divided by 100000000000 0000000000000000000000000000000000 (that's 1 followed by 43 zeros). Suffice it to say, this is an unimaginably small amount of time. And, to be clear, it's not that we necessarily *can* explain everything from the Planck Time on, but that we currently definitely *cannot* explain anything before it.

To sum up where we've gotten so far: there may have been a singularity. If there was, it was immediately followed by an era we can't really say much about called the Planck Time.

Truthfully, the whole timeline of the early universe is still very much an extrapolation and, I will readily admit, one that we shouldn't entirely trust. A universe that starts with a singularity and expands from there goes through an unimaginably extreme range of temperatures, from basically infinity at the singularity to the cool comfortable environment of the cosmos today, sitting at about 3 degrees above absolute zero. What we can do is make inferences about what physics would be like in all those environments, which is how we get the ordering I present in this chapter. And though the standard Big Bang theory of steady expansion from a singularity has some major problems (which we'll get to imminently), we can still learn a lot about how physics works by thinking about what might have happened if the standard theory is right.

---

* Named for Max Planck, an early originator of quantum theory. There's also a Planck energy, length, and mass, all defined through various combinations of fundamental constants, one of which is the Planck constant, which is central to anything with a quantum nature. If you find the Planck constant in your equations, you know things are liable to get weird.

THE GUT ERA

According to the standard Big Bang story, after the Planck Time comes the GUT era. (I'm using the term "era" here to refer to something that lasts about 10^{-35} seconds, and the term GUT to mean something unrelated to human anatomy.) "GUT" stands for Grand Unified Theory, which is the physics-utopian ideal of a "unified" theory that describes how all the forces in particle physics worked together under the extreme conditions of the universe at this early stage. Even though the universe was cooling rapidly, it was still so hot that the amount of energy at every point in space was over a trillion times higher than the energy generated by the most powerful collisions in our most advanced particle colliders. Unfortunately, and partly because of that factor of a trillion preventing us from doing experimental tests, the theory is currently still very much under construction. But for a theory we do not currently have, we can say quite a lot about it, and how it's different from what we see today.

In everyday life in the modern universe, each of the fundamental forces of nature has a distinct role. Gravity holds us to the ground, electricity keeps our lights on, magnetism holds our shopping list to the fridge, the weak nuclear force makes sure our backyard nuclear reactor keeps glowing a nice steady blue, and the strong nuclear force prevents our bodies' protons and neutrons from decomposing into their component parts. But the physical laws governing how these forces work, how they interact with each other, and even whether or not it's possible to tell them apart, depend on the conditions under which they're measured. Specifically, the ambient energy, or temperature. At high enough energies, the forces start to merge and combine, rearranging the structure of particle interactions and the laws of physics themselves.

It's been known for some time that, even under everyday

circumstances, electricity and magnetism are aspects of the same phenomenon, which is why electromagnets are a thing and why dynamos can generate electricity. This kind of unification is like candy to physicists. Any time we can take two complex phenomena and say, "Actually, when you look at it *this* way, they're THE SAME THING," we basically explode with physics joy. In some ways, this is the ultimate goal of theoretical physics: to find a way to take all the complicated messy stuff we see around us and rearrange it into something pretty and compact and simple that just *looks* complicated because of our weird low-energy perspective.

Where the forces of particle physics are concerned, this quest is called Grand Unification. Based on theory and extrapolations from what we see in lab experiments, it's thought that at very high energies, electromagnetism, the weak force, and the strong force all come together to be something else entirely, such that there's no way to distinguish them—they're all part of the same kind of big particle-energy mix governed by a Grand Unified Theory. There have been a few GUTs developed and proposed, but the difficulty of accessing the energy scales where the unification occurs makes them hard to confirm or rule out, so we'll just call this "an area of active research" that would very much appreciate any more funding you'd like to throw at it.

You might notice that gravity is not invited to the GUT party. To bring gravity into the picture, we need something grander and more unifying than even a Grand Unified Theory—we need a Theory of Everything (aka TOE). There's a general belief among physicists that sometime around the Planck Time, gravity *was* somehow unified with the other forces (along with the dragons or whatever else was happening then). But, as we discussed before, general relativity and particle physics don't like to work together in their current form, and so we've made even less progress toward a TOE than a GUT. Many people are placing their bets on string theory being the ultimate TOE. But if you thought GUTs were hard to verify experimentally, TOEs may actually be impossible to test, at least with anything like the technology we can currently conceive of. Arguments rage from time to time about whether or not this is the case, and whether or not untestable theories should even be called science. I don't think the situation is quite as dire as that. Cosmology may offer solutions (and no, I'm not just saying that because I'm a cosmologist). In certain cases, with a bit of creativity, there are some tantalizing possibilities for testing predictions of string theory and related ideas with observations of the cosmos. If we make it through a couple of apocalypses in the next few chapters, we'll see how cosmology might be able to show us more about the ultimate, tying-it-all-up-with-a-bow fundamental structure of the universe than any particle experiment can.

But let's get back to our story. We have just escaped the throes of Planck Time–quantum-gravity confusion and are enjoying the fundamental-force unity of the very slightly less speculative GUT era.

COSMIC INFLATION

What happened next is still a matter of debate, but the near-consensus in cosmology is that sometime around this moment the universe suddenly experienced the mother of all growth spurts—a process we call *cosmic inflation*. For reasons we're still trying to understand, the expansion of the universe suddenly went into very high gear, with the region that would someday become our entire observable universe increasing in size by a factor of more than 100 trillion trillion (i.e., 10^{26}). Of course, that only brought it up to about the size of a beach ball, but given that the starting point was unimaginable tininess smaller than any known particle, and the growth happened over the course of something like 10^{-34} seconds, we have reason to be impressed.

The theory of inflation came about to solve a few really perplexing problems in the standard Big Bang model. One has to do with the weird uniformity of the cosmic microwave background, another with its tiny imperfections.

The uniformity problem is that the standard Big Bang cosmology doesn't offer any explanation for how the whole observable universe, including parts on totally opposite sides of the sky, ended up being the same temperature at early times. When we look at the afterglow of the Big Bang, we see it as the same everywhere to extremely high precision, which is, when you think about it, a really strange coincidence. Normally, two things can become the same temperature if they're in what we call *thermodynamic equilibrium*. This just means there's a way for them to exchange heat, and time for them to do so. If you leave a cup of coffee in a room for long enough, the coffee and the air in the room will interact, and eventually you'll have room-temperature coffee and a very slightly warmer room. The problem with the standard picture of the early universe is that it doesn't include a situation in which two distant parts of the universe could interact and come to an agreement about a temperature. If we take two points on opposite sides of the sky and work out how far apart they are now, and how far apart they were at the very beginning, 13.8 billion years ago, it becomes clear that there was never a time in the history of the universe when they were close enough that light beams could have traveled back and forth between them to carry out the equilibrium process. A beam of light that started at one of those points at the beginning of the universe would not have had time even in 13.8 billion years to cover the distance necessary to get to the other. They are, and always have been, outside each other's horizons: unable to communicate in any way.* So either it's the most massive coincidence in the universe, or something happened early on to let the equilibrium happen.

The problem with the imperfections is a little simpler to articulate. It's just this question: where did those minuscule density fluctuations in the cosmic microwave background come from, and why are they patterned the way they are?

Cosmic inflation solves both these problems, along with a few others. The basic idea is that there was a time in the early universe, after the singularity but before the end of the primordial fireball stage, when the universe was expanding astonishingly fast. This helps by allowing for a period early on when a very small region could come into equilibrium, after which the rapid expansion would take that little nicely settled region and stretch it out to cover our entire observable universe. Imagine taking a complicated abstract painting and blowing it up so large that your whole view is just one color. The expansion essentially zoomed in on a part of the universe that was small enough to have already become the same temperature, and made the whole observable universe out of just that region.

Inflation also conveniently explains the density fluctuations,

---

* There's a subtlety here in this simplified explanation that has always bugged me. I'm telling you on the one hand that these regions have never communicated in the history of the cosmos, but I'm also telling you that the universe started with a singularity where, one might suppose, all the distances between things were zero. The reason this doesn't solve the problem is this. Take two points on opposite sides of the sky now. For the sake of argument, we'll say they were at zero distance at time zero. The problem is, at every time AFTER zero, those parts were not in contact—they couldn't have had any information exchange (like a light beam carrying information about temperature). What about, you say, zero itself? While we can label the first moment zero, it is literally zero time. Time began at the singularity. So there wasn't any time for the information exchange (because there wasn't any time) and every moment after that has the "too far apart to communicate" problem.

if we invoke a bit of quantum physics. The essential difference between the physics of the subatomic world and that of everyday life is that on the scale of individual particles, quantum mechanics imbues every interaction with an intrinsic, inescapable uncertainty. You may have heard of Heisenberg's Uncertainty Principle: it's the idea that there's a limit to the precision of any measurement, because the uncertainty built into quantum mechanics will always smear out the result in one way or another. If you measure a particle's position very precisely, you won't be able to determine its speed, and vice versa. Even if you just leave a particle alone, all of its properties will be subject to some amount of random shifting, and every time you measure it you could get a slightly different answer.

How does this tie into the cosmic microwave background? The hypothesis is that inflation was driven by a kind of energy field subject to quantum fluctuations: random jumps up and down. These fluctuations, which would normally only be ephemeral blips on a microscopic scale, change the density on the tiny scales where they happen, and then get stretched out to large enough regions to become substantial hills and valleys in the density distribution of the primordial gas. The little splotches we see in the cosmic microwave background make perfect sense if they are the natural evolution, over hundreds of thousands of years, of the fluctuations set down in the first 10^{-34} seconds of the cosmos. And those same little splotches are what eventually grew into all the galaxies and clusters of galaxies we see today.

The fact that the distribution of the largest structures in the universe can be exactly patterned on the minute wiggling of a quantum field is something that never fails to blow my mind. The links between cosmology and particle physics are never clearer, or more visually striking, than when we look at the cosmic microwave background.

But we're getting ahead of ourselves. The CMB is still, on

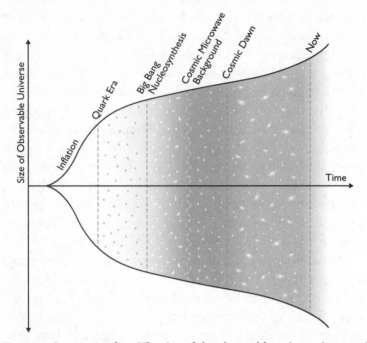

Figure 6: Cosmic timeline. The size of the observable universe increased rapidly during inflation, just after the very beginning. The universe has been expanding (at a slower rate) ever since. Labeled here are some of the important moments in the history of the cosmos.

these terms, eons away. We've only covered 10^{-34} seconds, and there's still plenty of story to tell.

When inflation ended, the super-stretched-out baby universe was left much colder and emptier than when it began. A process called "reheating" brought it back up to a high temperature everywhere, at which point the ordinary march of steady expansion and cooling carried on.

THE QUARK ERA

While the pre-inflation cosmos was likely ruled by a Grand Unified Theory, the post-inflation cosmos was moving closer to the laws of physics we see today. There was still a way to go, though. At this point, while the strong nuclear force had broken away from the GUT all-in-one particle physics party, electromagnetism and the weak nuclear force had not yet distinguished themselves; they were still somehow merged together as a single "electroweak" force. But particles were starting to emerge from the primordial soup—specifically, quarks and gluons.

Quarks, these days, are most commonly encountered as the components of protons and neutrons (which, together, are called hadrons). Gluons are the "glue" that binds quarks together via the strong nuclear force, and they are aptly named. They're so good at binding quarks together that while quarks have been found in twos or threes or even occasionally quads and quintuples, finding a single quark in isolation has so far proved impossible. It turns out that if you have two quarks bound together (in an exotic particle called a meson), you have to put in so much energy to separate the quarks that before you can get them apart, the energy you just expended spontaneously produces two more quarks. Congrats! Now you have two mesons.

In the very early universe, however, the usual rules didn't apply any more to singleton quarks than they did to anything else. Not only were the forces of nature operating under different laws, the universe contained a different mix of particles, and temperatures were so high that bound states of quarks could not exist in a stable form. Quarks and gluons bounced around freely in a hot roiling mix called a quark-gluon plasma—kind of analogous to the inside of a fire, but nuclear.

This "quark era" lasted until the universe reached the ripe old age of a microsecond. Meanwhile, somewhere in there (probably around the 0.1 nanosecond mark) the electroweak force split into electromagnetism and the weak nuclear force. Also around that time, something happened to create a distinction between matter and antimatter (matter's annihilation-happy evil twin), allowing most of the universe's antimatter to annihilate away.* Exactly how and why that happened is still a mystery, but as matter we can be glad it occurred, so we're not constantly running into antimatter particles and vanishing in a puff of gamma rays.

In contrast to the GUT era, we actually know quite a lot about the quark era and the quark-gluon plasma. The theory is pretty well developed and less of a departure from standard particle physics than a GUT, and experiments confirm the predictions we make when we start from electroweak theories and extrapolate from there. But the real coup is that we can actually *re-create* quark-gluon plasma in the lab. Particle colliders like the Relativistic Heavy Ion Collider (RHIC) and the Large Hadron Collider (LHC) can, by smashing together gold or lead nuclei at extremely high speeds, momentarily produce tiny fireballs so hot and dense that they smush together all the particles and momentarily fill the collider with a quark-gluon plasma state. By watching the debris "freeze" out into ordinary hadrons, scientists can study the properties of this exotic matter as well as the way the laws of physics act under those extreme conditions.

If seeing the CMB gives us a glimpse of the Big Bang, high-energy particle colliders are giving us a taste of the primordial soup.†

---

* Today, we find antimatter in certain kinds of particle reactions, but mostly notice it because when an antimatter particle meets its matched regular matter particle, they annihilate, destroying each other and creating a burst of energy.

† They are also, incidentally, giving us clues about the other end of time:

BIG BANG NUCLEOSYNTHESIS

After the quark-gluon plasma phase, the universe started to cool down enough for some familiar particles to form. At around a tenth of a millisecond, the first protons and neutrons formed, followed shortly by electrons, laying down the building blocks of ordinary matter. Somewhere around the two-minute mark, the universe cooled to a comfortable billion degrees Celsius, hotter than the center of the Sun but cool enough to allow the strong force to bring those shiny new protons and neutrons together. They formed the first bonded atomic nucleus: a form of hydrogen called deuterium (one proton bonded to one neutron; technically a single proton can be considered a nucleus too, since it's the center of a hydrogen atom). Soon, nuclei were forming left and right. Some fraction of the protons and neutrons started joining together to make helium nuclei, tritium, and a smattering of lithium and beryllium. This process, called Big Bang Nucleosynthesis, went on for about half an hour, until the universe cooled down and expanded enough that the particles were able to get away from each other instead of fusing together.

One of the great validations of the Big Bang theory is the fact that we find a close match between our observations in the cosmos and the calculated abundance of elements we expect from the Big Bang based on our estimates of the temperature and density of that primordial fireball. The agreement isn't perfect—there's some lingering confusion around the lithium abundance that may or may not be telling us about some extra weirdness going on in the early universe—but with hydrogen,

---

recent breakthroughs have shown us evidence that the end of the universe might come in a totally unexpected way, and it may happen at any moment. But that's all covered later in the book; let's not get ahead of ourselves. We'll probably make it to Chapter 6.

deuterium, and helium, measuring how much we actually see out there and comparing it to what we calculate *should* happen if you shove the entire cosmos into a nuclear furnace results in some absolutely beautiful concordance.

As an aside, the fact that pretty much all the hydrogen in the universe was produced in the first few minutes means that a pretty large fraction of what you and I are made of has been hanging around the universe in one form or another for almost as long as the universe has been here. You may have heard that "we are made of stardust" (or "star stuff" if you're Sagan), and this is absolutely true if we measure *by mass*. All the heavier elements in your body—oxygen, carbon, nitrogen, calcium, etc.—were produced later, either in the centers of stars or in stellar explosions. But hydrogen, while the lightest, is also the most abundant element in your body *by number*. So, yes, you hold within you the dust of ancient generations of stars. But you are also, to a very large fraction, built out of by-products of the actual Big Bang. Carl Sagan's larger statement still stands, and to an even greater degree: "We are a way for the cosmos to know itself."

THE SURFACE OF LAST SCATTERING

After Big Bang Nucleosynthesis, things in the infernoverse started to settle down, relatively speaking. At that point, the mix of particles was more or less stable, and would remain so until the time of the first stars, millions of years later. But for hundreds of thousands of years, the cosmos was still a hot, humming plasma composed mostly of hydrogen and helium nuclei and free electrons, with photons (particles of light) bouncing around between them.

Over time, the expansion of the universe gave all that radiation and matter room to spread out. I sometimes imagine experiencing this phase of the early universe like a journey from the center of the Sun outward, but instead of moving through space, you're moving through time. You start in the center of the Sun, where the heat and density are so high that atomic nuclei are fusing together to make new elements. The solar interior is opaque with light, with photons continually bouncing off electrons and protons so violently that it can take hundreds of thousands of years of constant scattering for a photon to reach the surface. Eventually, as you move outward, the plasma becomes less dense and light is able to travel farther between scatterings. At the surface, it can stream freely out into space.

In a similar way, a journey through time from the first few minutes of the universe to about 380,000 years later takes the entire cosmos from that hot dense plasma to a cooling gas of protons and electrons that can finally come together to make neutral atoms, allowing light to travel freely between them instead of constantly bouncing off the charged particles. We call the end of this fireball stage of the early universe the "surface of last scattering," because it's a kind of surface *in time* at which light goes from being trapped in plasma to traveling long distances across the cosmos.

This is what we see when we look at the cosmic microwave background: the moment that delineates the end of the Hot Big Bang, and the transition to a universe in which space is dark and silent and light travels through it. It is also the beginning of the cosmic Dark Ages—the time when the gas is slowly cooling and condensing into clumps, drawn in by the tiny density blips set up by those initial fluctuations. Sometime around the hundred-million-year mark, one of those clumps becomes so dense it is able to ignite into a star, and Cosmic Dawn has officially begun.

COSMIC DAWN

The transition from a dark, gaseous universe to one shimmering with the light of galaxies and stars was driven, largely, by a kind of matter so exotic that we haven't been able to re-create it even in the most powerful particle colliders. In the mix with the radiation, hydrogen gas, and a sprinkling of other primordial elements was a substance we know today as *dark matter*. It's not really dark, but rather invisible: seemingly unwilling to interact with light in any way. No emission of radiation, no absorption, no reflection. A light beam heading toward a clump of dark matter, as far as we can tell, passes right through. But where dark matter really shines* is in its ability to gravitate. When regular matter tries to condense in a clump drawn in by its own gravity, that matter has pressure, and pushes back. But dark matter can condense without feeling this force. A side effect of not interacting with light is not interacting much with anything at all, since, under most circumstances, collisions between particles of matter come from electrostatic repulsion, which needs interaction with light to take place. (Photons are particles of light, but they're also the carriers of electromagnetic force, so if something is invisible, it doesn't experience electromagnetic attraction or repulsion.) No electromagnetism, no pressure.

The first little blips of higher-density matter, set down by the fluctuations at the end of inflation, contained a mix of radiation, dark matter, and ordinary matter. Because the ordinary matter had pressure, and mixed with the radiation, at first only the dark matter was able to clump together due to gravity without immediately bouncing back. Later on, when the universe expanded more and the radiation streamed away from

---

* I'm very sorry.

the cooling matter, gas was able to fall into these gravitational wells and begin to condense as stars and galaxies. Even today, the structure of matter on the largest scales—the cosmic web of galaxies and clusters of galaxies—is scaffolded by a network of dark matter clumps and filaments. At Cosmic Dawn, those invisible clumps and filaments first started to light up, as stars and galaxies ignited and shone, sparkling along the network like fairy lights in the darkness.

THE ERA OF GALAXIES

The next big transition for the universe came when so much starlight was coursing through space that it was able to ionize the ambient gas that had become neutral at the end of the cosmic fireball stage. This intense starlight broke hydrogen atoms back apart into free electrons and protons, creating giant bubbles of ionized hydrogen gas surrounding the brightest collections of galaxies. Those bubbles expanding through the cosmos marked the Epoch of Reionization ("re-" because the gas had been ionized during the Big Bang at the beginning, and was now being ionized again by the stars). That transition, which was completed at about the billion-year mark, is now one of the frontiers of observational astronomy, and we are only just beginning to understand how and when it occurred. In the almost 13 billion years since that time, things have carried on in much the same way, with galaxies forming and combining, supermassive black holes building up mass in galactic centers, and new stars being born and living out their lives.

So, here we are. The cosmos as we see it today is a vast, beautiful web of galaxies shining in the darkness. Our own pretty blue-and-white world orbits a moderately sized yellow star in a galaxy that is, in every meaningful way, fairly close to the average. While we have yet to find clear signals, this unremarkable galaxy might be teeming with life, as the debris of long-exploded supernovae creates the basic ingredients of biology on each of the worlds scattered around a hundred billion stars. By current estimates, as many as one in ten star systems has a planet of just the right size and distance from its star to sustain liquid water on its surface—a hint, though not a certain one, that life could find a way to thrive. In the trillion other galaxies visible across the observable universe, there could be countless other species, with their own civilizations, arts, cultures, and scientific endeavors, all telling their stories of the universe from their own perspectives, slowly discovering their own primordial past. On each of those worlds, creatures like, or unlike, ourselves might be detecting the faint hum of the cosmic microwave background, deducing the existence of the Big Bang and the staggering knowledge that our shared cosmos does not go back forever in time, but had a first moment, a first particle, a first star.

And those other beings, like us, might be coming to the same realization: that a universe that is not static, that had a distinct beginning, must also, inevitably, have an end.

CHAPTER 3:

Big Crunch

Let's start with the end of the world, why don't we? Get it over with and move on to more interesting things.
N. K. Jemisin, *The Fifth Season*

On a dark, moonless night in autumn, in the Northern Hemisphere, look up and find the wide W shape of the constellation Cassiopeia. Stare at the space below it for a few seconds and, if the sky is dark enough, you'll see a faint fuzzy blur almost as wide across as the full Moon. That blur is the Andromeda Galaxy, a great spiral disk of about a trillion stars and a supermassive black hole, all of which are hurtling toward us at 110 kilometers per second.

In about four billion years, Andromeda and our own Milky Way galaxy will collide, creating a brilliant light show. Stars will be flung chaotically out of their orbits, forming stellar streams that stretch across the cosmos in graceful arcs. The sudden smashing together of galactic hydrogen will spark a minor explosion of star birth. Gas will ignite around the previously dormant central supermassive black holes, which will meet in the middle of everything and spiral into each other. Jets of intense radiation and high-energy particles will pierce the chaotic tangle of gas and stars, while the central region of the new Milkdromeda galaxy is irradiated with the X-ray-hot glow of a whirlpool of doomed matter falling into the new, even more supermassive black hole.

Even in the midst of this great galactic train wreck, the vast distances between stars will make head-on stellar impacts unlikely: the Solar System as a whole will probably survive, more or less. Not the Earth, though. By that time, the Sun will have already begun to swell to red-giant size, heating up the Earth enough to boil the oceans and completely sterilize the surface of all possibility of life. If, however, any outpost of human ingenuity manages to sustain a presence in the Solar System to watch, the combining of the two great spiral galaxies will be an awe-inspiring and beautiful process, playing out over billions of years. When the particle jets and supernova fires have calmed, the resulting mass will become a giant ellipsoidal collection of old and dying stars.

As cataclysmic as it may be to those in the midst of it, the merging of galaxies is an everyday occurrence in the cosmic sense, and strangely lovely if viewed from an extremely distant vantage point. Large galaxies tear apart and cannibalize smaller ones; adjacent stellar systems combine with one another. Our own Milky Way shows evidence of having consumed dozens of smaller neighbors—we can still see trails of stars tracing giant arcs around our own galactic disk like debris from an interstellar car crash.

Throughout the universe, however, collisions like this are becoming increasingly rare. The universe is expanding: space itself—that is to say the space between things, not the things in it—is getting bigger. This means isolated individual galaxies and groups of galaxies are getting, on average, farther and farther apart. Within each group and cluster, mergers can still occur. Our immediate neighborhood collection of stellar systems, members of the blandly named Local Group, are a ragtag gang of small and irregular galaxies dominated by the two giant spirals, and we are all destined to get nice and cozy sooner or later. Venture farther afield, though, beyond a few tens of millions of light-years, and everything appears to be spreading out.

The big question, in the long term, is: will this expansion continue indefinitely, or will it eventually stop, turn around, and bring absolutely everything crashing together? How do we know the expansion is even happening?

When you're in a universe that's expanding the same way in every direction, it doesn't look like expansion, per se, but rather like the odd phenomenon of everything else receding from you . . . wherever you are. From our perspective, we see distant galaxies all moving away from us, as though we emit some kind of repulsive force. But if we were suddenly in a galaxy a billion light-years from here, we would see the same phenomenon: the Milky Way and everything else beyond a certain distance would be receding from *that* point. This behavior is a somewhat counterintuitive consequence of space getting bigger in the same way, at the same rate, everywhere.

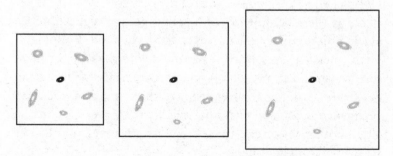

Figure 7: Illustration of cosmic expansion. Here, the increasing size of the universe at three different moments is represented by the increasing size of the square from left to right. As time goes on, galaxies move apart from one another, but they do not get bigger with the expansion of space.

The upshot is that every point in the universe is the nexus of what appears to be a powerful uniform repulsion. Technically, the universe doesn't have a center. But we're each the center of our own *observable* universe.* And from our perspective, all the galaxies farther out than our near neighbors are careening away from us as fast as they can. It's not us; it's cosmology.

Cosmic expansion was harder to discover than you might think. While galaxies other than our own have been visible through telescopes since the 1700s, their distances are so great and motions so slow (as judged by human timescales) that determining how they were moving relative to us, and even that they were galaxies at all, took more than two centuries. Even now, the most powerful telescopes can't see the motion directly—the galaxies don't appear to be farther away every time we look. But we can detect it by carefully teasing it out of a seemingly unrelated property of galaxies: the color of their light.

If you've ever heard the rising and suddenly falling "vroo-oom!" sound of a race car passing by, or the shift in tone when a siren approaches and retreats, you're familiar with the Doppler effect. A Doppler shift of the sort you normally experience is the phenomenon of a sound getting higher pitched as the object emitting it moves toward you, and dropping in pitch when it moves away. It has to do with the way the pressure waves in the air pile up on approach and stretch out on departure, changing the frequency of the sound that you hear. Frequency is, after all, just how quickly the waves hit you one after another. In sound, these are pressure waves, and higher frequency is higher pitch.

---

* Being the center of your own universe might sound appealing at first, until you factor in that the observational evidence for this is that everything is trying to get away from you as fast as possible.

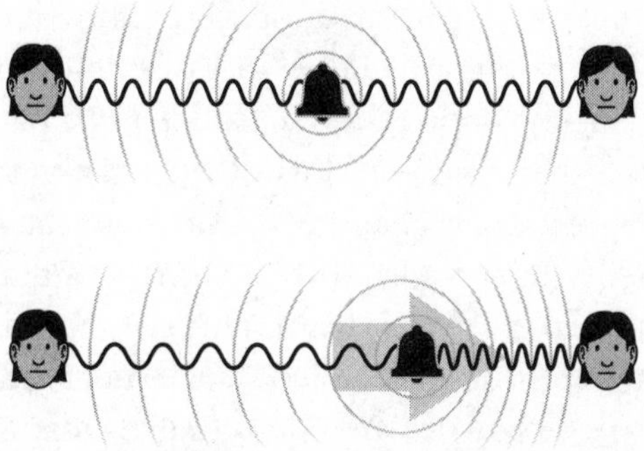

Figure 8: Illustration of a Doppler shift. When the source of the sound is stationary, the frequency heard by two stationary observers is the same. When the source of the sound is moving, the sound is stretched out to lower frequencies for the observer the source is receding from and compressed to higher frequencies for the observer the sound is approaching. The former hears a low note, while the latter hears a high one.

It turns out that light does something similar. A light moving toward you quickly will shift to a higher frequency, and one moving away will shift to a lower frequency. For light, frequency corresponds to color, so this shift looks like a color change. The electromagnetic spectrum extends far beyond the visible, but as a shorthand, when a Doppler shift of light occurs, a shift up is called a *blueshift* (because high-frequency visible light is on the blue end of the spectrum) and a shift down is called a *redshift*. Extremely blueshifted visible light could get all the way to gamma rays, while extremely redshifted light would show up as a radio signal. This phenomenon is one of the most important and versatile tools in astronomy, as it allows us to see, just from the color of a star or a galaxy, if it's moving toward us or away.

Of course, in practice, it's not quite that simple. (Astrophysics can be frustrating that way.) Some stars and galaxies are just redder, inherently, than others. How, then, to know if something is red because it's just red, or red because it's receding?* The key is that the light is never just one color, but a spread across frequencies—a spectrum. Characteristic patterns in the spectrum of a star are due to bits of the light being absorbed or emitted by different chemical elements in the star's atmosphere. When you spread out the light through a prism, different colors appear at different intensities, and dark lines or gaps show up at those specific frequencies where atoms in the star's atmosphere have absorbed the light—the light at those frequencies was removed by the gas before it could reach you. These features produce a kind of bar code unique to each element, with a pattern of lines that astronomers can recognize at a glance. So, for instance, light passing through a cloud of hydrogen will appear with a specific comblike pattern of dark lines when it's spread out across all the frequencies. We know from laboratory tests where the lines should be, and what the pattern should look like, and we can repeat the process with the patterns from other elements as well. If a star has a recognizable comblike pattern in its spectrum, but the lines appear at the "wrong" frequencies, that indicates the light from the star has been shifted by the star's motion. If each line is shifted in the same way to lower frequencies, that's a redshift and the star is moving away. If each line is shifted high, that's a blueshift and the star is approaching. And how far the lines have shifted tells us how quickly the star is moving.

Astronomers have gotten very good at this kind of measurement. Redshift or blueshift is now one of the most straightforward things to observe about any source of light in the universe, provided a spectrum is taken and it has any recognizable line patterns at all. We can use it to see how stars within our galaxy move relative to us, or to detect the tiny wobble

---

* We sometimes have a similar problem with "small" versus "far away."

of a star being pulled back and forth by the orbit of a planet around it.

And when it comes to distant galaxies, we can now use the redshift to measure not only how they're moving relative to us—toward us or away, and how quickly or slowly—but how far from us they are while they're doing it. How does that work? The expansion of the universe means that however a galaxy might be moving through its own space, the fact that the space between here and there is expanding means that it will also, in general, be moving away from us. And how quickly it's moving away from us depends on how far away it is now.

In 1929, astronomer Edwin Hubble was looking at galaxy redshifts when he noticed a striking, conveniently simple pattern. Galaxies that are farther away have, on average, higher redshifts. This relationship has allowed us to both confirm the expansion of the cosmos and map out its evolution. Translating redshifts to speeds, the pattern Hubble detected meant that the more distant a galaxy, the faster it's receding from us.

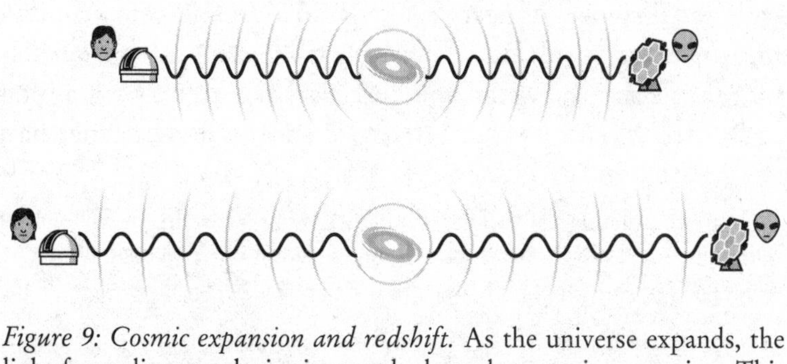

Figure 9: Cosmic expansion and redshift. As the universe expands, the light from distant galaxies is stretched out by cosmic expansion. This means that we will observe the light from a distant galaxy at a longer wavelength (redshifted) as the expansion of the universe proceeds. Because the expansion is happening everywhere, another observer watching a distant galaxy somewhere else in the cosmos will also see that galaxy's light redshifted.

Imagine stretching a slinky between your hands. (Just stretching, not bouncing. This is for science.) As you move your hands apart, each curl of the slinky moves only a finger's width away from the one next to it, but the two curls at opposite ends will end up a few feet apart in the same amount of time. If space is expanding uniformly in all directions, the same kind of relationship should hold, and that's exactly what Hubble's observations found. Mathematically, this gives us a conveniently simple rule of thumb: a galaxy's apparent speed is *directly proportional* to its distance. Meaning, first, more distant things are moving away faster. And second: there's a single number by which you can multiply any galaxy's distance to get its speed. While it was Hubble's data that ultimately proved the relationship and produced an estimate for that number, the proportionality was actually predicted theoretically a few years earlier by Belgian astronomer and priest Georges Lemaître. The relationship is accordingly known as the Hubble-Lemaître Law.* And the constant of proportionality (the number by which you multiply the distance) is called the Hubble Constant.

The crucial part for us here is the connection between redshift and distance. It means that we can look at a distant galaxy, measure the redshift, and determine from that exactly *how* distant the galaxy is. (With some technical caveats.)†

But redshift is *also* connected to cosmic time. The expansion

---

* It is frequently called just Hubble's Law by those in the astronomy community, but in 2018 the International Astronomical Union voted to officially recognize Lemaître for his contribution by adding him into the name. As a theorist myself, I approve.

† In the "nearby" universe, where the recession speeds are small, this is just a simple division problem: velocity divided by Hubble Constant equals distance. For more distant sources, it is complicated by the fact that the Hubble Constant isn't actually constant over all of cosmic time, and the proportionality isn't a strict proportionality when the speeds are very high. It is safe to assume in general that if anything in cosmology sounds extremely simple, it's an approximation, a special case, or the ultimate The-

of the universe makes a lot of things weird in astronomy, and one of them is that we use what is essentially a color, written as a number, to denote speed, distance, and "the age the universe was at the time when the thing was shining." Physics is wild.

Here's how it works. If we measure a galaxy's redshift, we know how quickly it's receding from us, and we can use the Hubble-Lemaître Law to get its distance. But because light takes time to travel to us, and we know light's speed, knowing the distance also tells us how long the light has been en route. That means that measuring a galaxy's redshift tells us how long ago the light left the galaxy. And since we know how old the universe is now, that tells us how old the universe was when that galaxy sent out the light we see.

Taking this all into account, astronomers can use redshift to refer to earlier epochs of the universe. "High redshift" is long ago when the universe was young; "low redshift" is more recent. Redshift 0 is the local, present-day universe; redshift 1 is seven billion years ago. At the high end, redshift 6 is a universe only a billion or so years into its life, and the very beginning of the universe, if we could see it, would have a redshift of infinity.

So: a high-redshift galaxy is a distant galaxy that existed when the universe was young, and a low-redshift galaxy is a relatively nearby object living in what is basically the "modern" universe.

The distance-age-redshift relationship is incredibly useful in cosmology. But it relies on the fact that the recession speed always increases with distance in a known way. What if the expansion were to suddenly slow down? What if it were to *stop*, and reverse course? One consequence is that it would completely throw off our distance-measurement rules

---

ory of Everything we've all been searching for all our lives. (I wouldn't bet on Option 3.)

of thumb and upset a lot of astronomers. Another, nearly as important consequence, depending on who you ask: it would spell doom for the universe and everything in it.

WHAT GOES UP

For as long as we've known that (1) the universe started with a Big Bang and (2) it is currently expanding, the logical next question has been whether it will turn around and come back on itself, ending in a catastrophic Big Crunch. Starting with some very basic and reasonable physics assumptions, there appear to be only three possibilities for the future of an expanding universe, and they are all fairly direct analogs to what can happen to a ball thrown into the air.

You're standing outside, on the planet Earth. You throw a baseball straight up. You have an inhumanly good arm, just for the sake of argument. (And air resistance isn't a thing.) What will happen?

In the usual case, the ball goes up for a while, responding to that initial push you've given it, but it starts being slowed in its ascent by the gravitational pull of the Earth as soon as it leaves your hand.* Eventually it slows so much that it stops dead in the air and reverses course, falling back toward you and the planet you're standing on. But if you were to throw the ball incredibly fast—specifically, 11.2 km/s, the escape velocity of the Earth—you could in principle give the ball so much of a push that it leaves the Earth entirely, slowing down slightly all the while, and only comes to rest infinitely far in the future (or, I suppose, when it hits something else). If you throw it

---

* Technically, the ball and the Earth are both pulling on each other, because gravity is a two-way street, but the amount of motion the Earth experiences due to the baseball's gravitational tug is . . . not much.

even faster, it'll be completely unbound from the Earth and just coast away forever.

The physics of an expanding universe follows very similar principles. There's the initial push (the Big Bang) that set off the expansion, and from that point onward the gravity of all the stuff in the universe (galaxies, stars, black holes, etc.) works against the expansion, trying to slow it down and pull everything back together again. Gravity is a very weak force—the weakest of all the forces of nature—but it's also infinite in range and always attractive, so even distant galaxies must pull toward each other. As in the baseball example, the question boils down to whether or not the initial push was enough to counteract all that gravity. We don't even have to know what the initial push was; if we measure the expansion speed now, and also measure the amount of matter in the universe, we can determine whether its gravity will be enough to make the expansion eventually stop. Alternatively, if we can infer the expansion speed in the distant past, we can determine how the expansion is evolving over time by comparing that number against the expansion speed today.*

If our universe *were* fated to someday suffer a Big Crunch, the first hint would be seen via just such an extrapolation. Before the collapse began, we'd be able to see that the expansion was faster in the past and had been slowing down, in a specific, doom-precipitating kind of way. Over time, with an increasing degree of certainty, we'd get signs of impending collapse many billions of years before it officially started.

But before we get into the data analysis, let's stop to ask what the transition to a contracting universe and eventual

---

* You might be wondering if we can just measure the expansion now and ten years from now, and see how it's changed. Unfortunately, our current technology doesn't allow for measurements this precise, but in the coming decades we might be able to make this comparison.

apocalypse would look like, once it gets going. That's really what you're here for, after all.

Right now, the more distant an object, the faster it recedes, and therefore the higher its redshift (the Hubble-Lemaître Law). In a collapse-fated universe, this pattern will continue right up until the expansion stops completely—that top-of-the-roller-coaster moment. But since the speed of light prevents us from seeing the entire universe at once, we will still *perceive* distant objects receding long after they start turning around in actuality. Even though in some global sense the most distant objects are barreling toward us more quickly than nearby ones, at first we see the opposite behavior. Every galaxy nearby, out to just beyond our cosmic neighborhood, will appear to slowly come toward us. As with the Andromeda Galaxy, its light will be blueshifted. Just beyond those, there will be a distance at which everything seems to be standing still, while more distant things are redshifted, seeming to recede. Over time, the blueshifted nearby galaxies approach faster and faster, and the standstill radius moves out. Soon, we all stop worrying about what's happening to distant objects as the rush of nearby galaxies into our region of space becomes impossible, or at least highly inadvisable, to ignore.

We might be slightly (if naively) reassured by the fact that we will have had some experience with such things by then: in this scenario, the first signs of collapse come long after our collision with Andromeda. Even with the most pessimistic estimates, any Big Crunch event can only occur many billions of years in the future—our universe has been around for 13.8 billion years and with respect to the possibility of future collapse, it is definitely no more than middle-aged.

As we already discussed, the Andromeda–Milky Way collision is unlikely to affect the Solar System directly. But the onset of universal collapse is another story entirely. At first, it might look fairly similar: galaxies colliding and rearranging, new stars and black holes igniting, some stellar systems flung off into space. Over time, though, it will become increasingly and terrifyingly clear that something very different is going on.

As galaxies get closer together and merge more frequently, galaxies across the sky will burst with the blue light of new stars, and giant jets of particles and radiation will rip through the intergalactic gas. New planets might be born along with those new stars, and perhaps some will have time to develop life, though the terrifying frequency of supernovae in this chaotic, collapsing universe might irradiate the new planets clean. The violence of the gravitational interactions between galaxies and between central supermassive black holes will increase, flinging stars out of their own galaxies to end up caught in the gravity of others. But even at this point, collisions of individual stars will be rare, and they will remain so until very late in the game. The destruction of stars comes about through another process, one that also ensures, with great finality, the destruction of any planetary life that might still be lingering on.

Here's how.

The expansion of the universe as it is occurring today does more than just stretch out the light of distant galaxies. It also stretches out and dilutes the afterglow of the Big Bang itself. One of the strongest pieces of evidence for the Big Bang, discussed in the previous chapter, is the fact that we can *actually see it*, simply by looking far enough away. What we see, specifically, is a dim glow, coming from all directions, of light produced in the universe's infancy. That dim glow is actually a direct view of parts of the universe that are so far away that, from our perspective, they are still *on fire*—they're still experiencing the hot early stage of the universe's existence, when every part of the cosmos was hot and dense and opaque with roiling plasma, like the inside of a star. The light from that long-burned-out fire has been traveling to us all this time, and, from sufficiently distant points, has just now arrived.

The reason we experience this as a low-energy, diffuse background (the cosmic microwave background) is that the expansion of the universe has stretched out and separated the individual photons to the point that they're now merely a bit of faint static. And the fact that they show up as microwaves is due to extreme redshifting. The expansion of the universe can do a lot, including taking the heat of an unimaginable inferno and diluting and stretching it out until it's just a faint microwave hum we might experience only as a tiny bit of static on an old-fashioned analog TV.

If the expansion of the universe reverses, this diffusion of radiation does too. Suddenly the cosmic microwave background, that innocuous low-energy buzz, is blueshifting, rapidly increasing in energy and intensity everywhere, and heading toward very uncomfortable levels.

But that's still not what kills the stars.

It turns out that there is something that can create more high-energy radiation than concentrating the afterglow from space itself being on fire. As the universe has evolved over time, it has taken what was, at the very beginning of the cosmos, a fairly uniform collection of gas and plasma and used gravity to collect that gas into stars and black holes.* Those stars have been shining for billions of years, sending their radiation out into the void to be dispersed, but not to disappear. Even the black holes have had their chance to shine, producing X-rays as the matter falling into them heats up and creates high-energy particle jets. The radiation produced by stars and black holes is even hotter than the final stages of the Big Bang, and when the universe recollapses, all *that* energy gets condensed too. So rather than being a nicely symmetric process of expansion and cooling followed by coalescence and heat-

---

* And other minor things like planets and people, but for the purpose of this discussion we can ignore those.

ing, the collapse is actually *much worse*. If you're ever asked to choose between being at a random point in space just after the Big Bang, or just before the Big Crunch, choose the former.* The collected radiation from stars and high-energy particle jets, when suddenly condensed and blueshifted to even higher energies by the collapse, will be so intense it will begin to *ignite the surfaces of stars* long before the stars themselves collide. Nuclear explosions tear through stellar atmospheres, ripping apart the stars and filling space with hot plasma.

At this point, things are really very bad. No planet that survived this long could possibly exist un-incinerated when stars themselves are exploded by background light. From here, the intensity of the universe's radiation becomes so high that it can be compared to the central regions of active galactic nuclei, the places where high-energy particles and gamma rays shoot away from supermassive black holes with so much power they make jets of radiation a thousand light-years long. What happens to matter in an environment like that, after it's reduced to its component particles, is uncertain. A collapsing universe will, in the final stages, reach densities and temperatures beyond what we can produce in a laboratory or describe with known particle theories. The interesting question becomes not "Will anything survive?" (because by this point it is very clear that the answer to that is a straightforward No), but "Can a collapsing universe bounce back and start again?"

Cyclic universes that go from Bang to Crunch and back again forever have a certain appeal in their tidiness. (And we'll explore these in more detail in Chapter 7.) Rather than a beginning from nothing and catastrophic, final end, a cycling universe can in principle bounce along in time arbitrarily far in each direction, with endless recycling and no waste.

Of course, like everything in the universe, it turns out to

---

* To quote the legendary D:Ream, "things can only get better."

be significantly more complicated. Based purely on Einstein's theory of gravity, general relativity, any universe with a sufficient amount of matter has a set trajectory. It starts with a singularity (an infinitely dense state of spacetime) and ends with a singularity. There isn't really a mechanism in general relativity to transition from an end-singularity to a beginning one, however. And there is reason to believe that none of our physical theories, general relativity included, can describe the conditions of anything close to that kind of density. We have a pretty good understanding of how gravity works on large scales, and for relatively (ha!) weak gravitational fields, but we have no idea how it works on extremely small scales. And the kinds of field strengths you'd encounter when the entire observable universe is collapsing into a subatomic dot are all *kinds* of incalculable. We can be fairly confident that for that particular situation, quantum mechanics should become important and do *something* to make a mess of things, but we honestly don't know what.

Another problem with a bouncing Crunch-Bang universe is the question of what makes it through the bounce. Does anything survive from one cycle to another? The asymmetry I mentioned between an expanding young universe and a collapsing old one, in terms of the radiation field, is actually potentially very problematic here, as it implies that the universe gets (in a precise, physically meaningful sense) messier with every cycle. That makes the cyclic universe less appealing from the standpoint of some very important physical principles that we'll discuss in later chapters, and it's certainly more difficult to fit into a nice neat reduce-reuse-recycle ecology.

THE ALLURE OF THE INVISIBLE

Bounce or no bounce, a universe with too much matter and not enough expansion is destined for a crunch, so checking where we're at in terms of that balance seems like a good idea. Unfortunately, measuring the matter content of the universe is complicated by the fact that not all matter is easy to see, and determining how much a galaxy weighs when all you have is a picture of it can be challenging at best. As far back as the 1930s, it was clear that just counting up galaxies and stars meant missing something important. Astronomer Fritz Zwicky studied the motions of galaxies moving around in galaxy clusters and noticed they seemed to be moving too quickly, and should by rights be flying off into empty space, like children on a merry-go-round that's spinning too fast. He suggested that perhaps there was some unseen "dark matter" holding everything together. That idea floated around the astronomical community as an unsettling possibility until sometime in the 1970s, when Vera Rubin came along and demonstrated once and for all that whole heaps of spiral galaxies really didn't make any sense without some extra invisible stuff.

Since Rubin's time, the evidence for dark matter has only strengthened, partly because we now know how important it was in the early universe, but it has remained stubbornly hard to directly detect, on account of being apparently uninterested in interacting with our particle detectors. The leading idea is that dark matter is some kind of as yet undiscovered fundamental particle that has mass (and therefore gravity) but doesn't have anything to do with electromagnetism or the strong nuclear force. Theories suggest it might be able to interact with other particles via the weak nuclear force, opening up some possibilities for detection, but the signal would be hard to find and we haven't yet seen it. What we *have* seen is a massive amount of evidence for its gravitational impact on stars and galaxies, and on the ability for stars and galaxies to form out of the primordial soup in the first place. And even better than that, we can see evidence for dark matter's existence in the shape of space itself.

One of Einstein's big insights (among many) was that gravity isn't best understood as a force between objects, but rather as the bending of space around anything that has mass. Imagine rolling a tennis ball across the surface of a trampoline. Now put a bowling ball in the center. The way the tennis ball falls toward the bowling ball, or curves as it goes past it, is a pretty good analogy for how objects move through space in the presence of large masses. The shape of the space itself is causing the object's trajectory to curve. But it's not just the paths of massive objects that are affected by the bending of space—even light responds to the shape of the space it's moving through. Just like a curved fiber-optic cable can make the light inside it turn a corner, a massive object bending space can cause light to curve around it. Galaxies and clusters of galaxies become distorting magnifying glasses for the objects behind them. Some of our most compelling evidence for dark matter comes from finding that this "gravitational lensing" effect is stronger than can be accounted for by the mass of the stuff we can actually see—some of the mass is due to something invisible. It turns out there's a *lot* of dark matter out there. The first attempts to weigh up the matter in the universe by looking at just the visible stuff gave us a wildly inaccurate accounting. Not long after Vera Rubin's studies, it became clear that the vast majority of the matter in the universe is dark.

But even when dark matter was properly accounted for, it was difficult to determine whether the density of matter in space was on one side or the other of the "critical density" that defined the border between a recollapsing universe and an eternally expanding one. Determining the contents of the universe was only one part of the problem; the other part was figuring out exactly how fast the universe is expanding, or, alternatively, how the expansion has changed over cosmic time. This, it turns out, is no mean feat.

In order to get a good measurement of the cosmic expansion rate for a reasonable fraction of the history of the universe, you need to survey a huge number of galaxies, at a range of distances. Then, for each galaxy, you need to work out two things: its speed and its actual physical distance from us. Astronomers worked out the *local* expansion rate with the Hubble-Lemaître Law back in 1929 (though the exact number for the proportionality was argued about for decades after, and is still a point of some controversy). But to answer the Big Crunch question, we need to know the expansion rate across a huge swath of cosmic time, which means a huge distance in space. That's not so much of a problem for the galaxy speed part of the equation, since this can be determined with redshift measurements, which are, generally, reasonably straightforward. Measuring *distance* accurately over billions of light-years, however, is a lot harder.

Studying the distances and speeds of galaxies using images from photographic plates in the late 1960s led astronomers to state with increasing confidence, despite quite a lot of remaining uncertainty, that we were, in fact, fated to collapse. This prompted a few astronomers to write some very exciting papers delving into exactly what that was going to be like. It was a heady time.

In the late 1990s, however, astronomers perfected a more precise method for measuring the expansion of the universe, involving stringing together several methods for measuring cosmic distance and applying them to extremely distant exploding stars. Finally, they could take the true measurement of the universe and determine, once and for all, its eventual fate. What they found shocked pretty much everyone, earned three of the team leads a Nobel Prize, and made a complete mess of our understanding of the fundamental workings of physics.

The fact that the discovery indicates that we are almost certainly safe from a fiery death in a Big Crunch has turned out to be cold comfort.* The alternative to recollapse is eternal expansion, which, like immortality, only sounds good until you really think about it. On the bright side, we're not doomed to perish in an apocalyptic cosmic inferno. On the, well, dark side, the most likely fate for our universe turns out to be, in its own way, much more upsetting.

---

* Recollapse isn't impossible, from our current understanding. If dark energy, which we'll discuss in the next chapter, has especially weird and unexpected properties, it could reverse our expansion. But the evidence so far doesn't seem to point us in that direction.

CHAPTER 4:

Heat Death

VALENTINE: The heat goes into the mix.

(He gestures to indicate the air in the room, in the universe.)

THOMASINA: Yes, we must hurry if we are going to dance.
Tom Stoppard, *Arcadia*

One of my earliest astronomy memories is of a 1995 *Discover Magazine* cover story proclaiming a "Crisis in the Cosmos." Something impossible was showing up in the data: the universe appeared to be younger than some of its own stars.

All the careful calculations of the age of the universe, based on extrapolating the current expansion back to the Big Bang, suggested the universe was somewhere in the vicinity of 10 or 12 billion years old, whereas measurements of the oldest stars in nearby ancient clusters gave a number closer to 15. Of course, estimating the ages of stars is not always an exact science, so there was a chance that better data might show that the stars were a bit younger than they looked, shaving maybe a billion or two years off the discrepancy. But extending the age of the universe to finish solving that problem would create an even bigger one. Making the universe older would have required scrapping the theory of cosmic inflation—one of the most important breakthroughs in the study of the early universe since the discovery of the Big Bang itself.

It would take another three years of combing through data, revising theories, and creating entirely new ways of measuring the cosmos before astronomers would find a solution that didn't break the early universe. It just broke everything else. In the end, the answer came down to a new kind of physics woven into the very fabric of the cosmos—one that would fundamentally change our view of the universe and completely rewrite its future.

MAPPING THE VIOLENT SKY

The scientists who discovered the solution to the cosmic age crisis in the late 1990s weren't trying to revolutionize physics. They were trying to answer a seemingly straightforward question: how quickly is the expansion of the universe *slowing down*? It was common knowledge at the time that the expansion of the cosmos was set off by the Big Bang, and that the gravity of everything inside it has been slowing it down ever since. Measuring one number—the so-called deceleration parameter—would tell us the balance between the outward momentum from the Big Bang and the inward pull of the gravity of everything the universe is made of. The higher the deceleration parameter, the harder gravity is pushing the brakes on cosmic expansion. A high number would indicate the universe is fated for a Big Crunch; a low one would suggest that even though the expansion is slowing, it will never completely stop.

Of course, to measure deceleration, you have to find a way to measure how quickly the universe was expanding in the past, and compare that to how quickly it's expanding now. Fortunately, that whole thing where we can see the past directly by looking at distant things, coupled with the bit where the expansion of the universe makes everything look like it's moving away from us, means that this is totally possible. All we have to do is look at something nearby, and something really far away, see how quickly they're each moving away from us, and apply a little math. Simple!

Okay, in practice it's not simple at all, because you have to know the distances as well as the redshifts, and distances are hard to measure across deep space. But suffice to say, the measurement is *possible*, if very, very difficult. Fortunately, astronomers have a vast and varied toolkit for measuring things in the cosmos, and in this case it turns out that cataclysmic thermonuclear explosions of distant stars do just the trick!

The short explanation is that certain types of supernovae make explosions whose properties are so predictable we can use them as mile-markers for the universe. They involve the violent deaths of white dwarf stars, which are, when they're not busy exploding, the kind of slowly cooling stellar remnant that our Sun will eventually become after it gets through its planet-murdering red giant phase. When a white dwarf grows to a certain critical mass (either by pulling matter off a companion star or by colliding with another white dwarf),* it detonates. This is called a Type Ia supernova, and it produces a kind of characteristic rising and falling of brightness and a telltale spectrum of light that we can pretty reliably distinguish from other cosmic conflagrations. In principle, if you understand the physics of this kind of explosion really well, you know how bright it would be up close, and factoring in how bright it looks from all the way out here, you can deduce how far the light has traveled. (We call this the "standard candle" method because it's like you have a light bulb where you

---

* Weirdly, as of this writing, we are still not actually sure which of these is the main mechanism by which this happens. We just see the star go off and we know at least one white dwarf was involved.

know the exact wattage. Once you have that information, you can always deduce the distance using the fact that the bulb will look dimmer when it's far away by a factor of the distance squared. Only we say "candle" instead of "light bulb" because it sounds more poetic that way.)

Once you have a measure of the distance, you need to know how fast the supernova is receding. For that, you can use the redshifting of the light from the galaxy the star exploded in, which tells you how quickly cosmic expansion is happening at that point. Use the distance and the speed of light to work out how long ago this whole thing went down, and you have a measurement of the expansion rate in the past.

In 1998, just a few years after that *Discover Magazine* article raising the alarm about the age of the cosmos, two independent research groups collecting observations of distant supernovae came to the same utterly unreasonable conclusion. That deceleration parameter—the one measuring how quickly the expansion rate was slowing down—was negative. The expansion wasn't slowing down at all. It was speeding up.

THE SHAPE OF THE COSMOS

If the cosmos were behaving itself, the basic physics involved in the expansion of the universe should be about as simple as throwing a ball up into the air, as we discussed in the previous chapter. Throw it too slowly, it goes up for a bit, slows down, stops, and falls down again: that's like a universe where there's enough matter (or a weak enough initial Big Bang expansion) that gravity wins and recollapses the universe. Throw the ball *incredibly inhumanly fast* and it might just escape the Earth's pull and drift out in space forever, always slowing: a universe perfectly balanced between expansion and gravity. Throwing it even faster than that means it'll escape and just coast forever, approaching a constant speed as the gravity of Earth becomes less and less of an influence: that's like a universe that keeps expanding forever, not having anywhere near enough matter in it to turn around the expansion or even slow down it very much.

Each of these possible universe types has a name and a particular geometry. The geometry isn't the external shape of the universe in the sense of it being a sphere or a cube or something. It's an internal property—something that can tell you how giant laser beams would behave while shooting across the cosmos on immense scales. (Because if you're going to measure a property of space, you may as well do it with giant laser beams.) A Big Crunch–fated universe is called a "closed" universe, because two parallel laser-cannon beams would eventually bend toward each other—it's the same kind of thing that happens to lines of longitude on a globe. What's happening in the cosmic case is that there's so much matter in a closed universe that *all of space* is curved inward. A perfectly balanced universe is "flat" because the beams would just stay parallel forever, in much the same way two parallel lines would stay parallel on a flat sheet of paper. A universe with way more expansion than gravity is called an "open" universe, and in that case, as you may have already guessed, the two laser beams would diverge from each other over time. The 2D-surface analog here is a saddle shape: try drawing parallel lines on a saddle (or, if you don't have a saddle handy, you can use a Pringles chip) and they'll get farther apart as they go. What these shapes represent is the "large-scale curvature" of the universe—the amount that space on the whole is distorted (or not) by the matter and energy within it.

The first thing all these possibilities have in common is that they all make sense, physically; they work well with Einstein's equations of gravity. The other is that for all of them, the present-day expansion is slowing down. At the time the supernova measurements were made, there was no reasonable physical

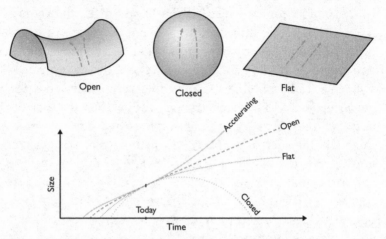

Figure 10: Open, closed, and flat universes and their evolution over time.
The diagrams above indicate the shape of space for three different cosmic models. In an open universe, parallel light beams diverge over time. In a closed universe, they converge. In a flat one, they remain parallel. The different geometries correspond to different fates of the cosmos as shown in the graph. In the closed case, there is enough gravity to cause the cosmos to recollapse, whereas in the open case, the expansion wins out and the universe expands forever. A perfectly balanced flat universe continues expanding but always slowing in its expansion rate. However, if a universe contains dark energy, its expansion can accelerate (while the geometry of space remains flat).

mechanism to make a universe *speed up* in its expansion. It was just as weird as if you were throwing a ball up into the air and it slowed down a bit and then suddenly *shot off into space for no reason*. Exactly that weird except for the ENTIRE UNIVERSE.

The measurements were checked, and rechecked, but they kept forcing physicists to the same conclusion. The expansion was accelerating.

These were desperate times, and they called for desperate measures. So desperate, in fact, that the astronomers invoked the existence of a vast cosmic energy field that could make the empty vacuum of space itself have an intrinsic outward push in all directions—a previously undetected property of spacetime that would make the universe expand forever, all on its own, from an ever-present energy source, never depleting. A *cosmological constant*.

NOT-SO-EMPTY SPACE

Unlike most monumental revisions of the foundations of physics, the cosmological constant was not a new idea at all. It was, in fact, originally Einstein's brainchild,* and it fit nicely into his equations of gravity governing the evolution of the universe. But it was based on a deeply mistaken notion, and by rights it should never have been written down in the first place.

Einstein's heart was in the right place. The purpose of the cosmological constant was to save the universe from catastrophic collapse. Or more accurately, from having catastrophically collapsed already. Being an expert, as he was, in all things gravitational, Einstein knew that all the data available pointed to the uncomfortable conclusion that gravity should have destroyed the universe long ago. This was 1917, half a century before widespread acceptance of the Big Bang theory, when the cosmos was still largely thought to be static and unchanging. Stars could live and die, matter might slightly rearrange, but space was *space*—it was just a background on which other things happened. So when Einstein saw that there were stars in the night sky, apparently stationary, he knew the universe was in trouble. Every one of those stars, he figured, should be gravitationally attracting every other star, and slowly draw-

---

* As much as the rest of us physicists might find it frustrating to admit, the guy had a lot of pretty good ideas.

ing together over time. It doesn't help for the other stars to be really distant, either; gravity is an infinite and purely attractive force. (It should be noted that this was a time before it was clear that other galaxies existed, or he would have applied the argument to galaxies instead. The problem would be the same.) In an unchanging universe, you can never be far enough away from something to not feel its pull, at some level, and that pull should be bringing you together over time. Einstein's own calculations said that any universe populated with massive objects should already have collapsed upon itself. The very existence of the cosmos was a contradiction.

This, of course, looked bad. Fortunately, Einstein found room in his General Theory of Relativity to add a little universe-rescuing tweak. Nothing in space could counter the gravity of the stars, but perhaps *space itself* could do it. Einstein had already developed a beautiful equation to describe how the shape of space responded to the gravitational attraction of all the stuff in the cosmos. All he had to do to ensure gravitational attraction wouldn't immediately collapse space was to decide his equation was incomplete, and to tack on a term that could stretch out the space between gravitating objects, perfectly balancing the contraction that gravity would otherwise cause. The term didn't represent a new component of the universe, but a property of space itself, where each piece of space has a kind of repulsive energy to it. When you have a lot of space and not much matter (like in the space between stars or galaxies), that repulsive energy can counter gravitational attraction.

Success! The equation worked. It nicely described a static universe in which the existence of other stars or galaxies doesn't immediately collapse the entire cosmos. Einstein had done it again.

Only problem: the universe is not static. This became apparent to the astronomical community a few years later, when it turned out that fuzzy smudges in the sky previously called "spiral nebulae" were actually other galaxies. Soon after, Hubble used the redshifting of those galaxies to show the universe was actually expanding. While a static universe with only attractive gravity is doomed, an expanding universe can be saved, at least temporarily, by its own expansion. The gravity might slow the expansion, and might eventually turn it around, but a universe can get along just fine for many billions of years on an initial growth spurt and the ongoing effects of that expansion. (How the expansion started is a whole other story, but all we need for this particular problem is for the universe to not be so profoundly doomed that it would *already* be toast, and either a cosmological constant or expansion can take care of that.)

The discovery of the expanding universe meant a whole new view of cosmology and a minor embarrassment for Einstein. He somewhat reluctantly removed the cosmological constant term from his equations and wandered off to try to revolutionize some other area of fundamental physics. And so things went, with the evolution of the universe making a reasonable amount of sense, right up until the supernova measurements made a mess of it all again in 1998. Accelerated expansion meant the cosmological constant had to be revived, with the only small mercy being that it was by then far too late for Einstein to say, "I told you so."

Just because a cosmological constant allows the universe to be accelerating in its expansion doesn't mean it's broadly considered to be a good and sensible solution.* There's nothing that explains why the cosmological constant term should have the value it does, from a theoretical standpoint. Why should it exist at all, except as a suspiciously convenient fix to our equations? And if we have to have a cosmological constant, why not a larger value? One of the most logically natural ways for

---

*You can tell you're in a demanding field when just SAVING THE UNIVERSE is not enough.

the universe to have a cosmological constant would be for the constant to derive from the vacuum energy of the universe—the energy of empty space that accounts for weird things like virtual particles that can quantum-fluctuate in and out of existence. But calculations of the vacuum energy required for quantum field theory give us a number 120 orders of magnitude larger than what the cosmological constant actually out there in space seems to be. In case you're not familiar with the term, an order of magnitude is a factor of 10. 100 is two orders of magnitude. 120 orders of magnitude is 10 raised to the power of 120. Even in astrophysics, where we sometimes play fast and loose with the numbers, this looks like a major discrepancy. So, if the cosmological constant isn't the vacuum energy quantum field theorists all know and love, what is it?

One suggested solution to this "cosmological constant problem" involves the hypothesis that the constant is small in our observable universe, but might take other values far away, and it's just a matter of chance that we are where we are. (Or, not chance, but necessity, if vastly different values of the cosmological constant would be hostile in some way to the development of life and intelligence, perhaps by making space expand too fast for galaxies to form.) Another possibility is that it's not a cosmological constant at all, but some kind of new cosmological-constant-mimicking energy field in the universe that might change over time, in which case there's a possibility that it evolved to what it is for some other reason.

Because we don't know whether it's really a cosmological constant or not, we generally call any hypothesized phenomenon that could make the universe accelerate in its expansion *dark energy*. To throw some more terminology into the mix, an evolving (i.e., nonconstant) dark energy is often called *quintessence*, after the "fifth element," a mysterious something-or-other that was popular to philosophize about in the Middle Ages and is not really much more precisely specified now. A nice thing about the quintessence hypothesis is that it could lead us to a theory with some parallels to the cosmic inflation at the beginning of time. We know that whatever it was that caused cosmic inflation eventually turned off, so perhaps a similar accelerated-expansion-causing field could have turned on since then, causing the acceleration we see today.

(One downside of the quintessence hypothesis is that it's theoretically possible for a dark energy that changes over time to violently destroy the universe. For instance, if whatever is accelerating the expansion now turns around, it could cause the universe to stop and recollapse, bringing us back to a Big Crunch after all. Fortunately, that looks very unlikely, though we can't entirely rule it out.)

In any case, based on observations at the moment, it really looks a lot like dark energy is a cosmological constant: an unchanging property of spacetime that has only recently (i.e., in the last few billion years) come to dominate the evolution of the universe. At early times, when the cosmos was more compact, there just wasn't enough *space* for the cosmological constant (which is a property of empty space) to do very much, so the expansion at that time was slowing down, just as we would have expected. But about five billion years ago, matter got so diffuse due to ordinary cosmic expansion that the inherent cosmological-constant-induced stretchiness of space started to really become noticeable. We can now measure the motion of supernova explosions so far away that they went off before the expansion started to speed up, meaning that we can trace out when the universe was decelerating, and pretty much exactly when it transitioned to acceleration. Dark energy still might be some new, dynamical field. But so far, a cosmological constant fits the data perfectly.

If we follow that through to its future consequences, it's kind of ironic, actually. Because now it seems that the term Einstein used to save the universe will end up spelling its doom.

THE INFINITE COSMIC TREADMILL

A cosmological-constant-induced apocalypse is a slow and agonizing one, marked by increasing isolation, inexorable decay, and an eons-long fade into darkness. In some sense, it doesn't end *the universe* exactly, but rather ends *everything in it*, and renders it null and void.

The reason a cosmological constant dooms the universe is that once it starts, the accelerated expansion never, ever stops.

The present-day observable universe is probably bigger than you think. The "observable" part refers to the region within our *particle horizon*. We define this as being the farthest we could possibly see, given the limitations of the speed of light and the age of the universe. Since light takes time to travel, and more distant objects are, from our perspective, farther in the past, there has to be a distance corresponding to the beginning of time itself. A distance at which, if a light beam started there at the first moment, it would take the entire age of the universe to reach us. This defines the particle horizon, and it's the farthest out we can observe anything at all, even in principle. Knowing that the universe is about 13.8 billion years old, logic would tell you that the particle horizon must be a sphere of radius 13.8 billion light-years. But that's assuming a static universe. In actual fact, since the universe has been expanding all that time, something just close enough to send its light to us 13.8 billion years ago is now much farther away — approximately 45 billion light-years. So we can define the observable universe to be a sphere of about 45 billion light-years in radius, centered on us.*

---

* If you were sitting in a different galaxy in a different part of the universe, you would also define your observable universe as being a sphere about

The closest we can get to seeing that "edge" is the cosmic microwave background, whose light comes from almost as far as the particle horizon. But a bit closer to us, we can also see ancient galaxies that are now more than 30 billion light-years away. The light we see from those galaxies started traveling through the universe long before they got to such incredible distances, though. If not, we wouldn't be able to see them at all, since the light coming from them now* can't ever reach us. It turns out that in a uniformly expanding universe, where the more distant things are receding more quickly, it is inevitable that there is a distance beyond which the apparent recession speed is faster than the speed of light, so light can't catch up.

"Wait!" you might be saying. "Nothing can travel faster than light!" This is a fair point, but it doesn't actually lead to a contradiction. While nothing can travel faster than light *through* space, there's no rule that limits how quickly things can happen to find themselves farther apart because they are sitting still in a space that's getting bigger between them.

The distance at which galaxies are currently moving away from us faster than light is surprisingly close, given how far we can actually see. We call it the Hubble radius, and it's around 14 billion light-years from here. I mentioned in Chapter 3 that we can label the distance to objects by their redshift factors—the amount that their light is shifted toward the red (low frequency/long wavelength) part of the spectrum due to the expansion of the universe. An object at the Hubble radius would have a redshift of about 1.5, meaning the light wave, and the universe itself, has stretched out to two and a half times its original length since the light was emitted.† But even that

_____

45 billion light-years in radius, centered on your own position. "Observable universe" is a subjective, literally self-centered, concept.

* As we have seen in Chapter 2, defining "now" is tricky.

†The relative-size-increase factor of the universe is 1 plus the redshift, so something nearby, at redshift 0, is in a universe whose size is the same as ours.

utterly unimaginable distance is, in cosmological terms, just around the corner. We've seen individual supernovae out to redshifts of almost 4. The most distant galaxies we've seen have been at redshift values of about 11, and the cosmic microwave background is at a redshift of around 1,100.

So how do we see so many things that are so far away that they're receding from us at more than the speed of light, and, in fact, always have been? If something is moving away at more than the speed of light, a light beam emitted from it is getting farther away from us, not closer. The trick is that the light we're picking up left the source long ago, when the universe was smaller and the expansion was actually slowing. So a light beam that started out being carried by the expansion of space away from us (even though it was emitted in our direction) eventually was able to "catch up" as the expansion slowed and it reached a part of the universe that was close enough for the recession speed to be less than the speed of light. It entered our Hubble radius from the outside.

Imagine you're standing in the middle of a very long treadmill that's going faster than you can run. Even running at your top speed, you're going to be dropping back. But if you don't get dragged back too far, and if the treadmill slows down enough, you can eventually make up the lost ground and start to move forward before falling off the back end. So if you're in a universe whose expansion is slowing down, you'll be able to see more and more distant objects as time goes on, as the light from distant objects catches up with the expansion. The "safe zone" in which the expansion speed is less than the speed of light, the Hubble radius, grows over time and envelops objects that were previously outside it. Our horizons,* so to speak, expand.

---

* The Hubble radius is not technically a horizon, in the physics sense of the word. The particle horizon is; it's a limit beyond which we cannot possibly obtain information about anything. The Hubble radius is just the radius at which the CURRENT expansion speed is the speed of light, but it

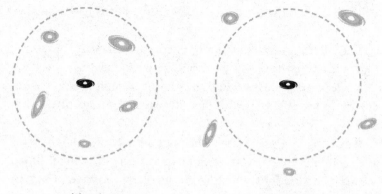

Now   Future

Figure 11: The Hubble radius now and in the future. As the expansion of the universe accelerates, galaxies that are currently inside our Hubble radius will be outside it. Eventually, no galaxies outside our Local Group will be visible.

Dark energy ruins everything, though. Because of dark energy, the expansion isn't slowing anymore—in fact, it has been speeding up for about the last five billion years. And while the Hubble radius is still technically growing in physical size, it's growing so slowly that the expansion is pulling previously visible objects outside it. We can see extremely distant objects whose light crossed into our Hubble radius before the acceleration began, but anything whose light isn't in the safe zone now will remain invisible forever. (More on that later.)

Even without the dark energy complication, an expanding universe can be a hard thing to wrap your head around.*

---

changes over time, and, as we've just discussed, objects can cross into it. People sometimes call it a horizon, but many cosmologists will get very worked up if you use that term.

* Not literally, obviously. That would be both impossible and extremely inadvisable.

The fact that the universe is expanding means it was smaller in the past: fine.

The fact that it was smaller in the past means that something that is far away now was closer in the past: okay.

That, in turn, means that there's a very distant galaxy we can currently see that was, billions of years ago, kind of nearby: right.

And long ago that galaxy shot out a beam of light that was originally moving directly away from us despite being pointed in our direction, but which from our perspective then sort of stopped and turned around and has just now arrived: sure, from a certain point of view, that might make sense.

BUT IT GETS EVEN STRANGER.

I'm sorry for shouting. I really am. But I'm not going to sugarcoat this. The universe is *frickin' weird* and this whole Hubble-radius-observable-universe thing is a big part of that and it makes deeply bizarre things happen. And now I'm going to tell you one of the most mind-blowing bits of weirdness I know about cosmology. You know how when something is far away, it looks smaller? This is a totally normal thing. The farther away something is, the smaller it looks. People look tiny from airplanes. Distant buildings can be covered with your thumb. Everyone knows that.

Except out there in the universe? Not so much.

For a while, sure, the more distant things are smaller. The Sun and the Moon look the same size to us because even though the Sun is vastly bigger, it's also a heck of a lot farther away. And for many billions of light-years, the more distant the galaxy is, the smaller it looks. As you would expect. But somewhere in the vicinity of the Hubble radius, that relationship *reverses*. Beyond that distance, the farther away something is, the larger it appears! This is super convenient for us astronomers, of course, as it allows us to see structure and details in galaxies that are extremely distant from us, and that in a sen-

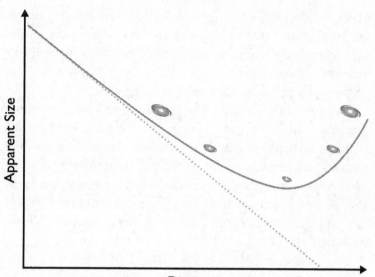

Figure 12: The apparent size of distant galaxies (assuming the same physical size) as a function of distance from us. Out to a certain distance, a faraway galaxy will appear to be smaller, but at some point, this turns around, and a more distant galaxy will look bigger in the sky. The dotted line indicates how the apparent size would relate to distance in a static universe.

sible universe would look like infinitesimal points. But if we think about it too much, it still seems like an utterly unreasonable way for geometry to work.

The reason for this reversal is related to the reason we can see things that are currently moving away from us faster than light. In the past, when the light was emitted, they were closer. So close, in fact, that they covered more of the sky. Even though they're much farther away now, the "snapshot" they've sent us has been traveling all that time, and is just reaching us now, showing us the ghostly image of a much closer thing. And the farther back in time you go, the smaller the universe was. So beyond a certain point, the balance between "the universe was smaller in the past" and "light takes a certain amount of time to get here" is such that a galaxy that is more distant than another galaxy *now* might have actually been *closer* when its light was emitted.

Look, I warned you it would be weird.

Anyway, if this is all deeply confusing and mind-boggling, that's totally okay and normal. Maybe try drawing some sketches on napkins, and then stretch out the napkins in every direction while on some kind of infinite treadmill running at an extreme speed over the course of billions of years, and hopefully it'll make sense then. Meanwhile, we should get back to what this all means for the future of existence. Because it really isn't good.

THE SLOW FADE TO BLACK

The assertion that "dark energy ruins everything" is not an overstatement. A universe whose expansion is accelerating is, paradoxically, one in which the influence exerted by the things in it is shrinking. Distant galaxies being dragged out of the Hubble radius by cosmic expansion will become lost to us. Galaxies whose distant past we can see now will slowly fade into darkness like ancient decaying photographs. In our own cosmic neighborhood, after the Milky Way and Andromeda merge, our little Local Group of galaxies will become more and more isolated, surrounded by darkness and the dying primordial light. All across the cosmos, invisible to us, other groups and clusters of galaxies will merge to form giant elliptical clumps of stars, burning brightly in the initial violence of the collisions but fading eventually to embers, whose glow will never reach beyond their own pool of expanding, emptying space.

Eventually, each new, dying supergalaxy will be utterly

alone. Nothing will again approach to bring in a fresh supply of gas to fuel new stars. The stars already shining will burn out, exploding as supernovae or, more often, sloughing off outer layers to become slow-burning relics, gradually cooling for billions or trillions of years. Black holes will grow, for a time. Some will engulf galaxies' worth of dead stellar remnants; some will stall in their growth, with no new matter approaching close enough to be consumed.

When the stars have all faded to darkness, the ultimate decay sets in.

Black holes begin to evaporate.

It was originally thought that black holes were eternal—capable of growing by consuming other matter but incapable of ever losing any mass. It makes sense that something defined by the fact that not even light can escape it would be a one-way, bottomless pit. But Stephen Hawking calculated in the 1970s that quantum effects on a black hole's horizon cause it to glow, faintly. The glow carries away energy—or, equivalently, mass—and the black hole shrinks. This process goes slowly at first, and then faster and brighter and hotter until a final explosion and disappearance at the end. Even the supermassive black holes at the centers of galaxies, with masses millions or billions of times that of the Sun, are destined to eventually fade and disappear.

Ordinary matter—the stuff making up stars and planets and gas and dust—suffers a similar, if less dramatic fate.

Most particles of matter are known to be, at some level, unstable. If left alone long enough, they decay into other things, dropping in mass and energy in the process. A neutron, for example, will eventually decay into a proton, an electron, and an antineutrino. While we've never seen a proton decay experimentally, we have reason to believe that can happen too, if you're willing to wait something like 10^{33} years. At that point even hydrogen atoms, which have been persisting as the most numerous atoms in the universe since the Big Bang itself, will finally cease to be.

The distant future of a universe governed by dark energy in the form of a cosmological constant is one of darkness, isolation, emptiness, and decay. But this slow fade is just the beginning of the ultimate end: the Heat Death.

The name "Heat Death" might sound like a misnomer for a state of the cosmos that is colder and darker than anything else in the history of creation. But in this case, the term "heat" is a technical physics term, not meaning "warmth" but rather "disordered motion of particles or energy." And it's not the death *of* heat, but a death *by* heat. It's the disorder in particular that kills us. Which is why we need to take a moment to talk about entropy.

Entropy is perhaps one of the most essential, versatile, and tragically obscure topics in all of science. It shows up everywhere—not just in the physics of everything from balloons to black holes, but also computer science, statistics, and even economics and neuroscience. Entropy is usually explained in terms of disorder. The more disordered a system, the higher its entropy. A pile of puzzle pieces has higher entropy than a completed puzzle; a scrambled egg has higher entropy than an intact one. In cases where "disorder" is not an immediately obvious property, you can think of entropy as a measure of how free or unconstrained the elements of the system are. To be concrete, a completed puzzle has low entropy because there's only one way for all the pieces to be arranged to make the puzzle whole, whereas a pile of pieces can be in any of a number of configurations and still successfully constitute a pile.

Though it's not so obvious in these examples, higher

entropy is also linked to higher temperature. This makes sense if you think of the difference between a block of ice and a cloud of steam. In order to be ice, the water molecules have to be arranged in a crystal structure, whereas the particles in steam are free to move around in three dimensions. But even just cooling the steam a bit reduces its entropy because the particles are moving less: they're more constrained, or less disordered.

The important thing about entropy, in cosmic terms, is that over time it goes up. The Second Law of Thermodynamics* states that in any isolated system, the total entropy can only increase, not decrease. In other words, order does not spontaneously appear out of nowhere, and if you leave something alone long enough, it will inevitably decay into disorder. Anyone who has tried to keep their desk tidy will understand this, the universe's most intuitive and maddening natural law.

Whether or not the universe itself counts as an isolated system can be a matter of some discussion, but taking it to be one leads us to the conclusion that the future of the cosmos is one of inevitably increasing disarray and decay. In fact, the Second Law is considered to be so inescapable and fundamental, it's been blamed for the passage of time itself.

The laws of physics generally have no regard for the direction of time; in most situations, reversing the equations in time makes no difference to the physics. The only part of physics that seems to care at all about which direction time is going is entropy. In fact, it's possible that the only reason we can

---

* The other laws are somewhat less exciting, though they do start with zero, just to be weird. Briefly, they are: 0) If one thing is in thermal equilibrium with another thing, and a third thing is in equilibrium with that, they're all in equilibrium with each other. 1) Energy is conserved and perpetual motion machines are impossible (sorry). 3) As something approaches absolute zero temperature, its entropy approaches a constant value.

remember the past and not the future is that "things can only get worse" is a truth so universal that it shapes reality as we know it.

"But wait!" you might say, "I completed the jigsaw puzzle! I created order! Did I just reverse the arrow of time?!"

Not exactly. The puzzle is not an isolated system, and neither are you. Technically, any local increase in entropy can be reversed with enough effort. It would be massively difficult, but you *could* unscramble an egg if you put in enough time and some incredibly sophisticated laboratory equipment. But the total entropy will always go up. In the case of the puzzle, the effort it takes you to put the pieces together requires an expenditure of energy, which means that you're breaking down food chemicals and releasing heat and waste products (like, you know, carbon dioxide) into your environment. That heats up the room, creates particulate waste, and probably wrinkles your shirt while you're at it. I don't know what an egg-unscrambling machine would do to its surroundings, but I'm pretty sure I wouldn't want to be in a closed room with it when it's running.

This, incidentally, is why leaving your refrigerator door open will ultimately make the whole kitchen hotter, and why air conditioners can contribute to global warming. Every attempt to bend some part of the world to our will creates disorder somewhere else, often in the form of heat.

As much as this has interesting applications for eggs and fridges and air conditioners, it all gets *much* weirder when we throw black holes into the mix.

Back in the 1970s, physicists were talking a lot about entropy and how the entropy of the *whole universe* must be increasing over time, and what the implications of that might be. At the same time, a young, not-quite-famous-yet Stephen Hawking and even younger postdoctoral researcher Jacob

Bekenstein were thinking about black holes and wondering if these bizarre, inescapable, spacetime garbage disposals might wreak havoc on the Second Law of Thermodynamics. What if, for example, you used your egg-unscrambler to unscramble the egg, and then pocketed the egg while throwing the whole messy hot unscrambler lab into the nearest black hole? Would you have decreased the overall entropy of the universe by putting the egg back together and getting rid of all the entropy you created in the process? After all, a black hole is defined as something that not even light can escape from, an object so massive and compact that its gravity turns outgoing light rays right around to send them diving back toward the central singularity. Beyond the event horizon of a black hole, the gravitational point of no return, nothing—not light, not information, not heat—can escape once it's gone in. Could hiding entropy behind black hole event horizons be the perfect crime?

Whatever other part of physics you have to break, don't bet against the Second Law of Thermodynamics. The solution to the entropy problem of black holes turned out to change everything we thought we knew about black holes and absolutely nothing about entropy. You can't hide entropy in black holes, because they have entropy of their own. Which means they have a temperature (they create heat). Which means they are not black at all.

What Bekenstein and Hawking eventually concluded about black holes is that a black hole has to have an entropy associated with it, to exist in accordance with the Second Law. Since that entropy should increase every time it swallows something, it makes sense that the entropy is connected to the size of the black hole itself—specifically, it's related to the total surface area of the event horizon. Throw a refrigerator into a black hole, and the mass increases by the mass of the refrigerator, which increases the horizon size and therefore the surface area.*

The fact that you can't have entropy without having a temperature means that black holes have to be radiating something (particles and radiation, specifically). And the only place they *can* radiate from is at or just outside the event horizon, since we still can't have anything coming out once it's gone in. So something weird has to be happening around there.

Fortunately, if we need weirdness in physics, we can always rely on the quantum realm to serve us up something good. In this case, Hawking made use of the quantum weirdness of *virtual particles*—pairs of positive- and negative-energy particles popping into and out of existence from the vacuum of space itself.† The idea is that this spacetime popcorn is happening *all the time*, everywhere, but usually it has no effect on anything because the two particles will appear and immediately annihilate against one another, both going back to being nothing again. But, Hawking said, near a black hole, you could have a situation where the negative-energy virtual particle falls past the horizon, leaving the positive-energy virtual particle so bereft that it becomes real and wanders away. The mass of the black hole would reduce a little as it absorbs that bit of negative energy, and the same amount of positive energy would appear to radiate off the black hole's horizon. Because these virtual particles are always popping up everywhere in space, any black hole that's not actively pulling in matter from its

---

* This isn't a tangible surface, but rather a sphere in space defined by the distance from the black hole's center to the Schwarzschild radius, which is what we call the distance from the singularity out to the horizon. The Schwarzschild radius is directly related to the black hole's mass.

†Real particles can't have negative energy, but these are virtual particles, which are just a totally different kind of beast, and are not to be confused with negatively *charged* particles like electrons.

environment should be gradually bleeding off mass through this evaporation process all the time.

As complicated as this might sound, it's still a vastly simplified picture, meant to capture just the basic idea without getting *too* technical, and it's an explanation that's used all the time. But it has never been particularly satisfying to me, since it seems to require the negative-energy particles to preferentially fall toward the hole, and the positive-energy ones to be traveling away from the black hole with enough energy to escape. It turns out that despite talking in these terms for popular audiences, Hawking never really wanted this explanation to be taken literally, and the real explanation involves calculating wave functions and the scattering that the waves experience in the vicinity of a black hole. I can't really get into it without a massive amount of math and a level of physics exposition that would probably require weekly lectures for two or three semesters, but I'm telling you about it because if it bugged me, it might bug you too, and I wanted to assure you that despite the inadequacy of the popular analogy the full calculation *does* make sense if you do it all rigorously, using general relativity and quantum field theory.

The point of this diversion was to say that we can safely assume that when facing the Heat Death, black holes do indeed evaporate away, leaving nothing but a bit of radiation spreading out through an increasingly empty universe. I hope that helps.

Also, aside from ultimately dooming all the black holes, the ability of horizons to radiate, and to account for the entropy of the things they contain, is actually an essential part of the Heat Death. Because our observable universe has a horizon too, and we're inside it.

MAXIMUM ENTROPY

A universe in the thrall of a cosmological constant is a universe that is evolving inexorably toward darkness and emptiness. As the expansion accelerates, there's more empty space, and thus more dark energy, causing more expansion, ad infinitum. Eventually, when the stars have burned out and the particles have decayed and the black holes have all evaporated, the universe is basically empty space with only a cosmological constant in it, expanding exponentially. We call this *de Sitter space*, and it evolves in the same way we think the very early cosmos did during inflation. Only, inflation eventually stopped. If dark energy really is a cosmological constant, the expansion can't stop and the cosmos will instead continue expanding, exponentially, forever.

So does a universe like this truly end, if it just keeps expanding? To answer that, we have to dig deeper into entropy, and the arrow of time.

Every time a star burns out or a particle decays or a black hole evaporates, it converts more matter into free radiation, which spreads through the universe as heat: pure disordered energy. Reducing something to heat radiation is dialing up its entropy to the maximum, because there are now no restrictions on the flow of energy. As the universe gets even emptier, that radiation gets more diluted, so you might think that the total entropy should drop as the temperature does. But that doesn't happen.

The way it works is this: when the universe reaches a state of steady exponential expansion, you can define a radius (from wherever you are) beyond which the rest of the cosmos is forever hidden. It's a true horizon in the sense that nothing beyond it could ever reach you. It turns out that this horizon, like a black hole's horizon, also has an entropy associated

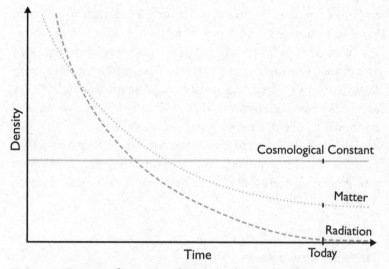

Figure 13: Density of matter, radiation, and cosmological constant over time. Because the density of dark energy (in the form of a cosmological constant) doesn't change as the universe expands, while everything else dilutes, it comes to dominate the energy density of the universe. Today, dark energy makes up around 70 percent of the universe, while matter is around 30 percent and radiation is a tiny amount.

with it, and thus a temperature. The difference is that instead of the heat going *out* like it does with a black hole, it goes *in*. The temperature is very small—something like 10^{-40} degrees above absolute zero—but when everything else has decayed, this radiation is all that's left to contain all the entropy of the universe. When the universe gets to this pure de Sitter state, it is a *maximum entropy universe*. From that point on, there is no way for the universe's total entropy to increase, which means, in a very real sense, the arrow of time is . . . gone.

I should just reiterate here that the arrow of time and the Second Law of Thermodynamics are so integral to the functioning of the universe that if there's no way for entropy to go up, *nothing can happen*. It is no longer possible for any organized structures to exist, for any evolution to happen, for any meaningful processes of any kind to occur. A necessary part of anything really happening is energy moving from one place to another. If entropy can't go up, then energy can't flow from one place to another without immediately flowing back, erasing anything that might have just, by chance, appeared to occur. Energy *gradients* are the basis of life, but also of any other structure or machine that performs any kind of work. Energy gradients can't exist in a universe that is just one giant (but very cold) heat bath. Heat is useless. Heat is death.

There are some caveats.

And to be clear, these are not caveats of the "well, technically there's this small detail" sort, but caveats of the "OMG this changes everything" sort.

This time, the weirdness comes down to a part of physics called *statistical mechanics*. This is what we use when we need to talk about something like temperature, which is really just the amount of motion in a system of particles, without painstakingly describing the path of each and every particle individually. Statistical mechanics is where the Second Law really shines, since it lets you describe a big complicated system in terms of one important property: its entropy. But it also introduces a kind of "out." That point about how it's an inescapable law of the universe that the entropy always increases? That technically only applies on average over sufficiently large scales. On the *quantum* scale, or even on large scales if you wait long enough, unpredictable fluctuations will, from time to time, spontaneously shift some part of the system into a lower-entropy state at random. The larger the system, the less likely it is that fluctuations could do much of anything at all, but in a universe that is in an eternal expansion and contains only a cosmological constant, there's a lot of time and space for waiting around, and even extremely low-probability events happen. It's *unlikely* that a whale and a bowl of petunias could suddenly pop into existence in completely empty space, but, in principle, if you wait long enough, it could happen.

This might come in handy. If anything can spontaneously pop into existence after the Heat Death, why not another universe?

The idea isn't as far-fetched as it sounds. There's a principle of statistical mechanics that says that any arrangement a system of particles finds itself in can happen again, if you wait long enough. Let's say you have a box filled with a gas of randomly moving molecules, and you take a snapshot of them at one moment and mark down what positions they're in. If you watch the box for a very long time, eventually you'll find the molecules in those positions again. The less likely the configuration, the longer it'll take, so a very rare event like all the particles huddling up in the lower right-hand corner of the box will take a lot longer to appear again, but in principle, it's just a matter of time. This is called *Poincaré recurrence*. If you have an infinite amount of time to work with, any state the system can be in is a state it WILL be in again, an infinite number of times, with a recurrence time determined by how rare or special that configuration is. In one rather arresting example, physicists Anthony Aguirre, Sean Carroll, and Matthew Johnson once calculated that if you were willing to wait something like a trillion trillion times the age of the universe, you could watch an entire piano spontaneously assemble itself in a seemingly empty box.

A post–Heat Death universe is, essentially, a very large, very slightly warmed box, with statistical mechanics stepping in to provide the random fluctuations. If the Big Bang is a state the universe has been in once, and the post–Heat Death universe is eternal (so eternal that, having lost the arrow of time, past and future are meaningless), there's no reason a Big Bang can't fluctuate out of the vacuum to start the universe anew.

Hold on, though. It gets even weirder. And more personal.

If *every* state the universe has ever been in could be revisited through random fluctuations, that means *this moment right now* could happen again, exactly the same in every detail. Not only could it happen again, it could happen again *infinitely many times*.

This possibility is of particular interest to cosmologist Andreas Albrecht, who has written about what he calls the *de Sitter Equilibrium* state. The basic idea of this equilibrium version of de Sitter space is that the origin of our universe and everything that happens in it can be thought of as the result of random fluctuations out of an eternally expanding universe containing only a cosmological constant. From time to time, a universe fluctuates out of the heat bath into a very low entropy starting state, and then evolves forward (with increasing entropy) until it gets to its own Heat Death, decaying back into the background de Sitter universe. And from time to time, the fluctuation doesn't produce a Big Bang, it just re-creates last Tuesday—specifically, that moment when you stubbed your toe on the kitchen table and spilled an entire cup of coffee on the floor. That moment. And every other moment of your life. And everyone else's.

If this sounds like a vaguely familiar image of dystopia, it's probably because it's disturbingly similar to a nightmare thought experiment first proposed by Friedrich Nietzsche in the late 1800s. In his book *The Gay Science* he writes the following:

> *What if some day or night a demon were to steal after you into your loneliest loneliness and say to you: "This life as you now live it and have lived it, you will have to live once more and innumerable times more; and there will be nothing new in it, but*

every pain and every joy and every thought and sigh and every-thing unutterably small or great in your life will have to return to you, all in the same succession and sequence—even this spider and this moonlight between the trees, and even this moment and I myself. The eternal hourglass of existence is turned upside down again and again, and you with it, speck of dust!"

Would you not throw yourself down and gnash your teeth and curse the demon who spoke thus? Or have you once experienced a tremendous moment when you would have answered him: "You are a god and never have I heard anything more divine." If this thought gained possession of you, it would change you as you are or perhaps crush you. The question in each and every thing, "Do you desire this once more and innumerable times more?" would lie upon your actions as the greatest weight. Or how well disposed would you have to become to yourself and to life to crave nothing more fervently than this ultimate eternal confirmation and seal?

Heavy.

For Nietzsche, the point of this proposal had nothing to do with thermodynamics and everything to do with an examination of the meaning, purpose, and experience of life as a human being. He likely never imagined such a scenario could be *literally, physically true*, as the de Sitter Equilibrium hypothesis proposes.

You could argue that these scenarios are not exactly the same thing. The quantum fluctuation that re-creates your experience of stubbing your toe might produce something that is exactly like you in every detail, but you, as an entity, would have been long dead by then. But this brings up questions of what it means to be *you*. Is the exact configuration of atoms you, or is there something ineffable and persistent about your consciousness that could never be re-created piece by piece? This is the same question that sparks heated debates among science fiction fans around teleportation and whether or not Captain Kirk was brutally murdered every time he stepped into a transporter beam, only to be replaced by a duplicate impostor that mistakenly believed itself to be him. We are unlikely to answer it here.

But it does bring up another wrinkle to the rebirth-by-quantum-fluctuation scenario—one that has as much to do with the transporter question as it does with the sperm whale and bowl of petunias, all wrapped up in a kind of quantum mechanical solipsism. It's a problem called Boltzmann Brains.

The idea is that if it's possible for the entire universe to quantum-mechanically fluctuate out of the vacuum, it's much *more* likely for just a single galaxy to do it, because a single galaxy is less complicated and requires less stuff to suddenly appear. And if it's more likely for a single galaxy to appear, it's more likely for a single solar system, or a single planet. In fact, far more likely even than that is that the only thing that fluctuates out of the vacuum is a single human brain, one that contains all your memories and is in the process of imagining that it lives on a perfectly functional world and is currently sitting in a coffee shop typing the words to the fourth chapter of a book about the end of the cosmos.

The Boltzmann Brain problem is the assertion that this unfortunate brain, doomed to quantum-fluctuate back into the vacuum almost instantaneously after its creation, is so vastly more likely to occur than a whole universe that, if we want to use random fluctuations to build our universe, we have to accept that we're much more likely to be just imagining the whole thing.

This question is not yet settled. Despite being one of the first people to propose the Boltzmann Brain problem in this context, Albrecht now comes down on the side that it's more likely for a de Sitter universe to create a very low-entropy state like the Big Bang than something small on the brink of reabsorption. The basic argument is that creating a low-entropy state might seem to take a lot of quantum fluctuation energy, but actually takes out only a little bit of total entropy from the system. Many cosmologists take the opposite approach and say that it's easier to fluctuate to a still relatively high-entropy state than to create a pocket where the entropy is very, very low. Settling this question could give us a handle on one scenario for the origin of the entire cosmos, as well as lend us some peace of mind with respect to our possible fate of infinitely playing back our most cringe-worthy moments forever.

And for some cosmologists, understanding how we started with a low-entropy state in the early universe, and determining once and for all whether or not we have to worry about Boltzmann Brains or Poincaré recurrences, are questions that shake the very foundations of our cosmological model. Trying to find a way to set up a low-entropy initial state has prompted some to hypothesize entirely new cosmic histories (as we'll discuss in Chapter 7), though the issue is very far from resolved. And the possibility of fluctuations is so disturbing to our picture of a sensible cosmos that it has been described by Sean Carroll as "cognitively unstable." It's not that it can't be true, but that if it is, nothing makes sense, and we might as well give up on trying to understand the universe at all. The jury is still out on this one.

If you're not too disturbed by the possibility of disembodied sentient brains popping into and out of existence, the possibility of rare random fluctuations can, in some sense, drag some order out of the Heat Death's nihilistic disarray. But even in this most optimistic view, a universe dominated by a cosmological constant unquestionably spells doom for any beings living within it, as absolutely every coherent structure is destined for dark, lonely emptiness and decay. Before dark energy was discovered, physicists like Freeman Dyson came up with speculative proposals that a machine whose computation constantly slows can persist for an arbitrarily long time into the cosmic future.* But even this ideal machine would be subject to entropic erosion via the Second Law, and would eventually disintegrate into waste heat in the face of the de Sitter horizon. The timescales for the achievement of maximum entropy—the true and timeless Heat Death—depend on estimates of the decay time of the proton, which are still uncertain. Nonetheless, we probably have a good 10^{1000} years or so before we and all other thinking structures fade from the possibility of memory.

It could be worse.

As dark energy goes, a nice, steady, predictable cosmological constant is something of a best-case scenario. Other possibilities are not ruled out, and one of them, phantom dark energy, leads to something more dramatic, more immediate, and, in a sense, much more final: the Big Rip.

---

*You may recognize Dyson's name from the sci-fi concept of a "Dyson sphere"—a monumentally huge sphere built around a star to capture 100 percent of its radiation for the purposes of powering an advanced alien civilization. Observational surveys for Dyson spheres, which look for the waste heat expected to be emitted by them in the infrared, have so far come up empty.

CHAPTER 5:

Big Rip

I keep thinking about this river somewhere, with the water moving really fast. And these two people in the water, trying to hold onto each other, holding on as hard as they can, but in the end it's just too much. The current's too strong. They've got to let go, drift apart. That's how it is with us.

Kazuo Ishiguro, *Never Let Me Go*

For a cosmic phenomenon that is arguably the most important thing in the universe, dark energy is surprisingly difficult to study. As far as we can tell, it exists everywhere in the universe, completely uniformly, woven into the fabric of space itself, and its only effect is to stretch space out so gradually that it has no detectable impact on any scale smaller than the vast expanses between distant galaxies. Dark matter physicists have it much easier—despite being just as invisible as dark energy, dark matter makes its presence very known by clumping around virtually every galaxy or cluster of galaxies we've ever seen, dominating the gravitational field, bending light, and altering the course of cosmic history from the very beginning. Dark energy, on the other hand, just . . . expands.

This doesn't completely prevent us from studying it. There are essentially two handles we have on dark energy: the expansion history of the universe and the way that galaxies and clusters of galaxies have grown over time. For both of these, we're peering into the distance and the past, tracing out the evolution of the cosmos over time. But no matter how we look, we're trying to tease out small effects using faint signals and statistics.

As challenging as these kinds of studies are, it's worth putting in the effort, since dark energy is both the dominant component of the cosmos and a sure sign of some new physics beyond our current understanding.

That, and the fact that depending on what dark energy turns out to be, it might violently and inescapably destroy the universe, much sooner than anyone ever imagined. Why wait for the slow fade of a Heat Death, if you can have a dark energy apocalypse as sudden and dramatic as the appropriately named Big Rip? Not only would it be a kind of destruction from which there is no escape, quantum-mechanical fluctuation or not, it would be one that could tear apart the very fabric of reality, rendering any thinking creatures in the cosmos helpless as they watch their universe being ripped open around them.

This alarming possibility is hardly an outlandish fringe idea. In fact, the best cosmological data we have not only fails to rule it out, but, from some perspectives, slightly prefers it. So it's worth spending a bit of time exploring what, exactly, it would do to us.

A COSMOLOGICAL NONCONSTANT

Dark energy is often assumed to be a cosmological constant that stretches space out, accelerating cosmic expansion by imbuing the universe with some inherent inclination for swelling. On large scales, this is a pretty good description. But within galaxies, solar systems, or in the close vicinity of organized matter generally, a cosmological constant has no effect. It can be more properly thought of as a force for isolation—if two galaxies are already distant from one another, they get more distant, and individual galaxies, clusters, or groups of galaxies find themselves more and more alone as time goes on. They also form a bit more slowly in the presence of a cosmological constant than they otherwise would. What the cosmological constant *cannot* do is break apart anything that is already, in any sense, a coherent structure. *What therefore gravity hath joined together, let not a cosmological constant put asunder.*

The reason for this small mercy of the cosmological constant (which, to be fair, does still destroy the whole universe eventually) lies in the "constant" part of the story. If dark energy is a cosmological constant, its defining feature is that the density of dark energy in any given part of space is constant over time, even as space expands. The expansion rate isn't constant, just the density of the stuff itself, in any given volume of space. This makes sense in a way, if every bit of space is automatically assigned a set amount of dark energy within it, but it's still *super weird*, because it means that as space gets bigger, the amount of dark energy increases to keep the density constant. It also means that if you draw a sphere of a given size anywhere in the universe and measure the amount of dark energy inside the sphere, and then do the same at some future time, you'll always get the same number, regardless of how much the outside universe has expanded in the meantime. If your original sphere contains a cluster of galaxies and some quantity of dark energy, in a billion years the amount of dark energy in that region will still be the same, so if it wasn't enough to mess up the galaxy cluster before, it won't be in the future. The balance between matter and dark energy in that sphere does not significantly change even as the rest of the cosmos seems to inexorably empty out.

This is reassuring. If you happen to be a clump of matter in the universe, and you would like to form a nice stable gravitationally bound galaxy, you can rest assured that once you get enough matter together to build something, dark energy won't ruin all that hard work.

Unless, that is, the dark energy is something more powerful than a cosmological constant.

As we discussed in the previous chapter, a cosmological constant is just one possibility for dark energy. All we really know about dark energy is that it's something that makes the universe expand faster. Or, more precisely, it has *negative pressure*. Negative pressure is a weird concept, because normally one thinks of pressure as something that pushes outward. But in Einstein's general-relativistic way of thinking about the universe, pressure is just another kind of energy, like mass, or radiation, and is thus gravitationally attractive. And in general relativity, gravitational attraction is just a consequence of the bending of space.

Remember the picture of the bowling ball creating a dent in the trampoline as an analogy for the effect of matter on the curvature of space? If you take general relativity into account, the dent is deeper if the ball is more massive, but also if it is hot, or if it has high internal pressure. So pressure, like other forms of energy, acts a lot like mass. From a gravitational perspective, pressure pulls. When you calculate the gravitational effect of a clump of gas, for instance, you have to factor in not just its mass, but its pressure, and both contribute to the gravitational impact that gas has on the stuff around it. In fact, the pressure contributes more to the spacetime curvature than the mass does.

What does that mean for something with *negative* pressure? If the pressure of some weird substance can be negative, it means that it can effectively *cancel out* the mass of the stuff, at least as regards its impact on the curving of spacetime. If you write down the pressure and density of dark energy in the form of a cosmological constant, in the appropriate units, the pressure is exactly the negative of the density.

We usually talk about the relationship between a substance's density and its pressure using a number called the *equation of state* parameter, written as w—it's equal to pressure divided by energy density, in units in which that comparison makes sense. Here, we're interested in the equation of state of dark energy, which, given enough time, will be the equation of state of the whole universe, since dark energy becomes more and more important in the expanding universe as everything else dilutes away. If the measured value of w = -1 exactly, that tells you that the pressure and the density are exactly opposite, and dark energy is a cosmological constant. Since the energy density in a cosmological constant is always positive, at first glance it seems as though it should act just like matter and amp up the gravity that slows down the expansion of the universe. But because the negative pressure is given a heavier weight in the equations, all a cosmological constant ends up doing is contributing toward accelerating cosmic expansion.

At least it does so in a predictable way. A cosmological constant, with w = -1, has a total energy density that is exactly constant over time as the universe expands, without increasing or decreasing. For dark energy with any other value of w, this is no longer the case. So it's important to figure out what we're really dealing with here.

In the years after dark energy was first discovered, it was clear that *something* was making the expansion of the universe accelerate, which meant there had to be something out there with negative pressure. It turns out that anything that has a value of w less than -1/3 gives you both negative pressure and accelerated expansion. But knowing the value of w could tell us whether dark energy is a true cosmological constant (w = -1 always), or some kind of dynamical dark energy whose influence on the universe might change over time. So astronomers went looking for a way to determine the value of w exactly. If dark energy turned out not to be a cosmological constant, this would indicate that we had not only discovered a new kind of physics acting on the universe, but one with the added bonus of being something even Einstein hadn't foreseen.*

For a few years, this was the name of the game: measure w, find out what's going on with dark energy. Measurements were made, papers were written, plots were drawn showing which values of w agreed with the data. The cosmological constant case looked like it just might win out.

But in the late 1990s and early 2000s, a small group of cosmologists pointed out a major undiscussed assumption their colleagues were putting into their calculations. It was a perfectly reasonable assumption to make, because neglecting it would violate certain long-held principles of theoretical physics so fundamental that no one wanted to upset them. But these principles weren't required by the data, and in the end, as scientists, our first loyalty has to be to the data. Even if it means rewriting the fate of the universe.

OFF THE EDGE OF THE MAP

The simple question physicist Robert Caldwell and his colleagues asked was: what if w is less than -1? Say, -1.5? Or -2? Up until this point, such a possibility was generally thought too outlandish to be considered. Plots in papers showing the "allowed" region for w based on the data tended to abruptly cut off at -1. The axis might go from -1 to 0, or -1 to 0.5, but -1 was a hard wall, the same way you might put a hard wall at 0 when guessing a person's height.

But when Caldwell looked at the problem, all the observations of w pointed to a value of -1 or something very close to it. Which suggested that there might be values below -1 that were also allowed by the data, if only someone were to check.

---

* He has to be wrong about *something*.

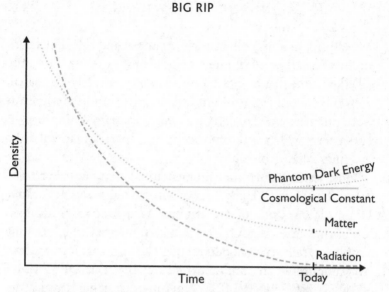

Figure 14: Evolution of dark energy in the form of a cosmological constant or in the form of phantom dark energy, compared with matter and radiation. While a cosmological constant keeps a constant density as the universe expands, in the case of phantom dark energy, the density increases.

This hypothetical dark energy with w less than -1 was dubbed by Caldwell "phantom dark energy" and would be deeply inconsistent with the aforementioned Important Theoretical Principles—specifically, the "dominant energy condition," which says, roughly, that energy can't flow faster than light.* This seems like a completely sensible condition to place on the universe, but it's subtly different than the usual statement that light (or any kind of matter) has an ultimate speed limit, and it's currently less of a proven physical principle than a Very Good Idea. Maybe it's flexible?

* In explaining the adoption of the term "phantom" in the first paper to touch on this idea in 1999, Caldwell wrote, "A phantom is something which is apparent to the sight or other senses but has no corporeal existence—an appropriate description for a form of energy necessarily described by unorthodox physics."

Caldwell and his colleagues went ahead and calculated constraints based on a full range of possibilities for w. Not only did they find that values below -1 were perfectly consistent with the data, they also found, through a simple, straightforward calculation, that if w is even infinitesimally lower than -1, dark energy will tear the entire universe apart, and it will do so in a finite, calculable time.

I just want to pause for a moment to say that this paper, titled "Phantom Energy: Dark Energy with w < -1 Causes a Cosmic Doomsday," is one of my absolute favorite papers in physics. It's not often that you get to take some very mild-seeming alteration to the current perspective, shifting a parameter down by a minuscule amount, and find out that this DESTROYS THE ENTIRE UNIVERSE. Not only that, it gives you a way to calculate exactly *how* the universe will be destroyed, and when, and what it will look like when it all goes down.

Which is as follows.

THE BIG RIP

You can think of it as an unraveling.

The first things to go are the largest, most tenuously bound. Giant clusters of galaxies, in which groups of hundreds or thousands of galaxies flow lazily around each other in long intertwined paths, begin to find that those paths are growing longer. The wide spaces traversed by the galaxies over millions or billions of years widen even more, causing the galaxies at the fringes to slowly drift away into the growing cosmic voids. Soon, even the densest galaxy clusters find themselves inexorably dissipated, their component galaxies no longer feeling any central pull.

From a vantage point within our own galaxy, the loss of the clusters should be the first ominous sign that the Big Rip is in progress. But the speed of light delays this clue until we are already feeling the effects much closer to home. As our local cluster, Virgo, begins to dissipate, its previously languid motion away from the Milky Way begins to pick up speed. This effect is subtle, though. The next one is not.

We already have astronomical all-sky surveys that are capable of measuring the positions and motions of billions of stars within our own galaxy.* As the Big Rip approaches, we start to notice that the stars on the edges of the galaxy are not coming around in their expected orbits, but instead drifting away like guests at a party at the end of the evening. Soon after, our night sky begins to darken, as the great Milky Way swath across the sky fades. The galaxy is evaporating.

From this point, the destruction picks up its pace. We begin to find that the orbits of the planets are not what they should be, but are instead slowly spiraling outward. Just months before the end, after we've lost the outer planets to the great and growing blackness, the Earth drifts away from the Sun, and the Moon from the Earth. We too enter the darkness, alone.

The calm of this new solitude doesn't last.

At this point, any structure still intact is straining under the push of the expanding space within it. The Earth's atmosphere thins, from the top. Tectonic motions within the Earth respond chaotically to the shifting gravitational forces. With only hours to go, the Earth cannot hold together: our planet explodes.

Even the destruction of Earth, could, in principle, be survivable, if, having interpreted the signs, you have already retreated to some compact space-based capsule.† But that reprieve is short-lived. Before long, the electromagnetic forces that hold together your atoms and molecules cannot hold up against the

---

* The newest one, called Gaia, is producing spectacularly detailed maps of the stars in our galaxy and is already giving us incredible insights into our cosmic history. What it will tell us about our fate is to be determined.

† When the danger is space itself, you want to be in a structure that has as little space in it as possible.

ever-expanding space within all matter. In the last tiny fraction of a second, molecules crack open, and any thinking beings still holding on are destroyed, torn atom-from-atom from within.

Beyond that point, there is no possibility of watching the destruction, but it carries on nonetheless. Nuclei themselves, the ultra-dense matter in the centers of atoms, are the next to go. The impossibly dense cores of black holes are eviscerated. And at the final instant, the fabric of space itself is ripped apart.

Unfortunately, we may never be able to say with certainty that we are safe from a Big Rip. The problem is that the difference between a Heat Death–fated universe and one headed for a Big Rip might literally be unmeasurable. If dark energy is a cosmological constant, the equation of state parameter w equals -1 exactly, and we get a Heat Death. If w is *at all* lower than -1,

Time from now:	Event
$\gtrsim$188 billion years	Big Rip
Time before Big Rip:	
2 billion years	Erase galaxy clusters
140 million years	Destroy Milky Way
7 months	Unbind Solar System
1 hour	Earth explodes
10^{-19} seconds	Dissociate atoms

Figure 15: Big Rip timeline (based on current worst-case scenario for w), adapted from Caldwell, Kamionkowski, Weinberg, 2003. The time until the Big Rip is at least about 188 billion years. The table indicates other moments of destruction in terms of approximately how long before the Big Rip they would occur.

even one part in a billion billions, dark energy is phantom dark energy, capable of tearing the universe apart. Because it's impossible to ever measure anything with complete, uncertainty-free precision, the best we may ever be able to do is say that if the Big Rip does occur, it will be so far in the future that all structure in the cosmos will have decayed already by the time it happens. Because even with phantom dark energy, the closer w gets to -1, the farther into the future the Big Rip is pushed. The last time I calculated the earliest possible Big Rip, based on the 2018 data release from the Planck satellite, I got something in the vicinity of 200 billion years.

Phew.

But given the potential consequences, both for the universe and for the structure of physics itself, we in the astronomical community put a pretty high priority on figuring out where we currently sit on the scale from w = -1 to Violent Cosmic Doom.* We can't measure w directly, but we can determine it indirectly by measuring the past expansion rate of the universe and comparing it to our best theoretical modeling of what different kinds of dark energy would have done. We glossed over this a bit in the previous chapter, but it turns out that even just determining the past expansion rate is far more difficult than it seems like it has any right to be. In principle, there are several ways to get at w, and some of them can be done in subtle ways that don't require calculating the expansion rate at specific distances. But the most straightforward way to get a handle on dark energy is to figure out our full expansion history. And it turns out all the weirdnesses of cosmology come crashing together if you try to do something as simple as answer the question, "How far away is that galaxy?"

---

* If you ask them, my colleagues will claim that their real motivation is understanding the nature of dark energy because of what it tells us about fundamental physics and our cosmological model. But *I* know it's really the dread.

LADDER TO HEAVEN

In order to meaningfully compare the local space-expansion rates at two distant points in the universe, you first have to know exactly *how* distant each one is. This is no big deal for something on Earth, or even something as close as the Moon, since you can measure the distance by bouncing a laser beam off of it and seeing how long the light takes to come back.* On those kinds of scales, the universe is pretty reasonable. It acts basically like an unchanging space where the distance from A to B is straightforwardly measurable and makes sense and everything works. When it comes to things outside the Solar System, it gets trickier, both because things that are more distant are harder to measure, and because on larger and larger scales, the expansion starts to change the definition of distance itself.

Astronomers have, over the years, patched together with duct tape and twine a set of overlapping definitions and measurements of distance that build upon one another. As kludgy as it still sometimes seems, it's the result of decades of innovations in observational astronomy and data analysis, and has given us an intuitive but frustratingly difficult-to-implement strategy known as the *distance ladder*.

Let's say you need to measure the length of a large room, and all you have is an ordinary-sized ruler. You could lay the ruler down repeatedly until you cover the length of the room, if you don't mind crawling around on the floor. Or you could be a bit more creative and measure the length of your stride, then just walk across the room, counting steps. If you chose

---

* Yes, we do this. It's called laser ranging, and the only reason we CAN do it is because the Apollo astronauts left a mirror up there. It's a handy tool for both seeing how far away the Moon is (fun fact: it's drifting away from the Earth at almost four centimeters per year) but also for testing how gravity works, by watching the orbit very, very carefully.

the steps method, you're creating a distance ladder: a system of measuring a large distance by calibrating your measurements with something more manageable.

In astronomy, the distance ladder has a series of rungs that allow it to extend out to objects that are billions of light-years away. Within the Solar System, direct laser measurements, orbital scalings, and even eclipses help us gather distance data. Beyond that, the next step is to use parallax. This is a method that takes into account the fact that when you change your vantage point, nearby things seem to shift their positions relative to a fixed background more than distant ones do. It's the same effect that makes a finger held in front of your face seem to jump back and forth when you close one eye and then the other. If we look at a nearby star in June, and then the same star in December, the fact that the Earth is in a different location in its orbit around the Sun means that the star will appear to have moved slightly with respect to more distant background objects. The closer it is, the bigger the shift. Unfortunately, for anything outside our own galaxy, these apparent motions are too small to be perceived, and we need another method—a way to determine the distance of bright objects just from the properties of their light.

The key to everything from here on out is the concept of a standard candle, which we discussed briefly in the previous chapter. This is a kind of object (such as a star) that has some physical attribute that tells you its brightness. Then, by seeing how bright it *looks*, you can tell how far away it is. Kind of like having a light bulb with "60 Watts" written on it. You know how bright it should be, but you'll get less light from it when it's far away.

Of course, nothing in space has its brightness helpfully stamped on it. But we have something almost as good. The breakthrough discovery that first allowed us to use standard candles in astronomy was due to the astronomer Henrietta

Swan Leavitt in the early 1900s.* Working at Harvard Observatory, she discovered that a certain kind of star known as a "Cepheid variable" brightened and dimmed in a predictable way. A Cepheid that's intrinsically brighter does slow, gradual pulsations, getting a little bit brighter and a little bit dimmer over a long period. A Cepheid that's intrinsically dimmer pulsates more quickly, with wide swings between its brightest and dimmest states.†

This discovery was revolutionary, and perhaps one of the most important in the history of astronomy, in that it let us finally measure the scale of the universe around us. It meant that anywhere a Cepheid could be seen, we could get a reliable distance and start to make a usable map. By measuring how quickly a Cepheid pulsed, and how bright it looked from here, Leavitt could tell you with great precision how bright it really was, and thus how distant.

How far does this get us? We can see Cepheid variable stars throughout the Milky Way and in nearby galaxies, so we can use parallax for the nearby ones, carefully calibrate the pulsation relationship, and then use the more distant ones to tell us the distances to other galaxies.

The next step in the distance ladder is a crucial one, but it's also where things get really messy, in every sense of the term. In the previous chapter, we mentioned that a certain kind of supernova can be used to measure distances. This kind of

---

* She wasn't referred to as an astronomer at the time. She was one of a group of women called "computers," who were hired to examine astronomical plates as cheap labor and who ended up doing a huge number of foundational calculations in astrophysics. Edwin Hubble, who used her discovery to measure the size and expansion of the universe, later said she deserved a Nobel Prize. Unfortunately, beyond being known and respected by her immediate colleagues, she was almost entirely unacknowledged in her lifetime.

† I like to think of the bright Cepheids like giant lazy Saint Bernard dogs, while the dim ones are excitable jumpy Chihuahuas.

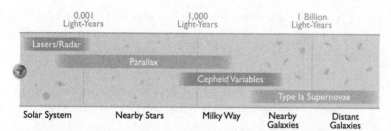

Figure 16: Cosmic distance ladder. For bodies within the Solar System, we can use lasers or radar (in addition to relationships between orbital times and distances) to measure distances. Distances to nearby stars can be measured with parallax, and Cepheid variable stars can help us determine distances within the Milky Way and some nearby galaxies. For more distant sources, we can use Type Ia supernovae.

explosion, a Type Ia supernova, is what happens when a white dwarf star somehow picks up some mass from another, equally unfortunate star and spectacularly rips itself apart. Because all white dwarf stars are fairly simple objects,* and because the explosion is governed by physics that we feel we have a somewhat decent handle on, Type Ia's were considered for a time to be good standard candles—the explosions all looked pretty similar. But it was later found that they're better described as standard*izable*, in the same way Cepheid variables are. If you can measure how the explosion peaks and dims, you can get a good sense for the total amount of energy put out by the explosion, and thus an idea of how bright it really is.

STAR LIGHT THERMONUCLEAR BRIGHT

But this book is about destruction, and I would be remiss if I glossed over Type Ia supernovae as just "a kind of exploding

---

* Simple for stars, anyway.

star." A white dwarf star, the kind of star our Sun is fated to eventually become, is itself a marvel of stellar evolution. And when one explodes, it does so by undergoing an all-out full-body thermonuclear detonation that outshines its entire galaxy.

If you are any kind of star, no matter what stage you are at in your life cycle, your existence depends on a careful balancing act between the pressure produced in your core and the gravity of the material you're made of. (We call this "hydrostatic equilibrium" but it really just comes down to the idea that the push out has to be equal to the pull in for a star to neither blow up nor collapse.) Most of the time, a star creates outward pressure by doing fusion reactions in its core—pressing nuclei together so tightly that they fuse and become a heavier kind of atom. For all the lightest elements, fusing them together produces radiation, and that radiation is the pressure that holds the star up against collapse.

For a star like the Sun, the outward pressure is provided by the fusion of hydrogen into helium. Most stars are, in fact, just giant helium factories, taking the abundant hydrogen in the universe and sticking it together, countless billions of times per second.

Let's consider the Sun, in particular, for sentimental reasons.

Right now, the Sun is happily burning hydrogen, creating a surplus of helium in its core and causing the temperature and pressure to change over time as the hydrogen-helium balance tips. Because the efficiency of the factory depends on both temperature and pressure, the energy output and size of the Sun will change over time—most noticeably, the Sun will get a little bit more radiant and a little bit bigger* over the next few million years.

---

* Based on current estimates, the Sun is already increasing in radius at about an inch per year. But at the same time, the Earth is expanding its orbit, such that we're moving away from the Sun at about 15 centimeters per year (I make no apology for the mixed units here), so it's not like the surface of the Sun is getting closer to us at the moment.

At somewhere around a billion years, we get to the part where we're all fried. But even after the Earth is well on its way to becoming a charred lifeless rock, the Sun has a long way to go yet. As that increased heat is incinerating the inner planets (Mercury and Venus) and evaporating all the oceans off the Earth, so much hydrogen will be burned off that there will only be a shell of hydrogen burning around the central helium-filled core. The core then gets hot enough to start fusing helium into oxygen and carbon and turns the Sun into a huge bloated red giant star. When the Sun eventually runs out of all hydrogen to fuse, a few billion years into its red giant phase, it'll begin its death throes in earnest. The core will start filling with oxygen, then carbon, the production fueled by the squeezing of the core by the gravity of the rest of the star. In the end, though, after the Sun has swelled to engulf the orbit of Venus and the Earth is a smoking ruin, the Sun's gravity won't be enough to maintain the temperatures needed for any further fusion. The outer atmosphere of the star will slough off, and the core will begin to contract.

You might think this would be the end of the Sun—depleted, transformed, and planet-devouring, left with no fusion reactions available that are strong enough to hold it up. Fortunately, there's a kind of pressure even stronger than fusion reactions that can keep the post-red-giant Sun and other stars like it from collapsing entirely, allowing it instead to live out its convalescence as a white dwarf star. And this pressure comes directly from quantum mechanics.

A QUANTUM HEAP

The first thing you need to know is that most of the subatomic particles you know and love—electrons, protons, neutrons, neutrinos, quarks—are fermions, which, in this context, means they are fiercely independent, in a particle physics kind of way. Specifically, they obey the *Pauli exclusion principle*, which says that they won't abide being in the same place and the same energy state at the same time. This is why, if you recall high school chemistry lessons, electrons attached to atoms end up in different kinds of "orbitals," which are really just energy states.

Anyway, in the core of a burned-out, collapsing star, there are so many atoms, pressed so tightly together, their electrons start to get antsy. At those kinds of pressures, the electrons aren't bound to specific atoms, but rather are packed in together in a big atomic mess so crowded that they have to jump to higher and higher energy states to keep from all being in the same one. This creates a kind of pressure, called *electron degeneracy pressure*, which is strong enough to halt the collapse of the star and create an entirely new kind of object: a white dwarf.

A white dwarf is a kind of star that isn't burning at all. It has no fusion. It is a solid object held up entirely by the quantum mechanical principle that electrons just don't like each other that much. And it can persist, silently smoldering, for billions and billions of years, until it slowly fades and cools and darkens, and is disintegrated in the Heat Death of the universe, ignited in the Big Crunch, or torn apart by phantom dark energy in the Big Rip, along with everything else.

Unless it gets just a little bit more mass.

Electron degeneracy pressure can do a lot. It can support an ENTIRE STAR. But only up to a point. If something happens to push the white dwarf past this point—it pulls in material from a companion star, or collides with another white dwarf—it will have too much mass for the degeneracy pressure to balance further collapse. Once that balance is tipped, a number of things happen in rapid succession.

The central core temperature of the star increases. Carbon begins burning. The material of the star starts to roil and churn, dragging more material in and out of the central flames. A deflagration tears through the star, creating a thermonuclear explosion so powerful that it rips the star apart, spectacularly and completely.

The explosion of a white dwarf star is so bright that it can briefly outshine its entire galaxy, and is visible to observers with telescopes billions of light-years away. Supernovae in distant parts of the Milky Way and nearby galaxies have even been seen without instruments, in ancient times, by the naked eye, *in the daytime.**

It's a matter of some frustration in the astronomical community that, aside from this broad-brush-stroke picture, we still don't know *exactly* how Type Ia supernovae happen. There are ongoing debates about whether they're primarily caused by material falling onto the white dwarf from companion stars or by white dwarf collisions. Simulating the explosion ripping through the star is also computationally extremely difficult. Most simulations result in incredibly impressive visualizations of bubbling, churning stellar material without actually quite getting to the exploding part. But they're working on it. (Stars, it turns out, are complicated. Especially when quantum mechanics and thermonuclear explosions are both important.)

The thing that makes us think we can learn anything useful from Type Ia supernova observations is the fact that we can reasonably expect that white dwarfs† are pretty much always at the same mass when they go off. In 1930, a twenty-year-old prodigy physicist from India named Subrahmanyan Chandrasekhar was on a ship traveling to England to begin his studies at Cambridge when he casually revolutionized stellar evolution in his

---

* Supernova 1006, seen between April 30 and May 1, 1006, was likely a Type Ia supernova caused by the collision of two white dwarf stars about 7,000 light-years away, in our own galaxy. The remnant, which in astronomical images looks a lot like a colorful ball of smoke, can still be seen today.

† They're "dwarfs," not "dwarves," for reasons that are not entirely clear.

free time. By improving on existing calculations and adding in important effects from relativity, he discovered a hard limit on the mass of any star held up by electron degeneracy pressure. That limit, about 1.4 times the mass of the Sun, became known, appropriately, as the Chandrasekhar Limit. Any white dwarf that gains enough mass to exceed that limit is immediately doomed to explode spectacularly as a supernova. And now that we know that the physics of the explosion is always the same, we know how bright a Type Ia supernova is intrinsically, and can therefore figure out its distance.

When Chandrasekhar's ship finally reached the shore, his breakthrough tore through the scientific establishment like a detonation front of knowledge, forever changing our view of these weird and wonderful explosive stellar objects. (Though not everyone was convinced. Apparently, celebrated big-shot astronomer Sir Arthur Eddington,* whose work Chandrasekhar had refined, was not pleased about being outshone by this upstart, and made the young physicist's life miserable for years before eventually giving in to superior calculational excellence.)

COSMIC POPCORN

The idea that all white dwarf stars explode when they gather enough mass to exceed the Chandrasekhar Limit gives astronomers hope that we can use these stars as distance benchmarks,

---

* If Eddington's name sounds familiar, it might be because he made an eclipse expedition in 1919 that provided some of the first observational evidence for Einstein's general theory of relativity. The observation of stars whose light grazed past the Sun en route to us showed that the light was being bent by the Sun's distortion of space. (This is the kind of observation that can only happen when the Sun is eclipsed.) A famous headline at the time proclaimed, "LIGHTS ALL ASKEW IN THE HEAVENS—MEN OF SCIENCE MORE OR LESS AGOG OVER RESULTS OF ECLIPSE OBSERVATIONS." Women of science were, presumably, unimpressed.

with some tweaks to account for slight differences in stellar circumstances.

Exactly how well we can do this is still a matter of incredibly intense debate in the astrophysics community. Which is understandable, as the stakes couldn't be higher. Type Ia supernovae are the gold standard* for distance measurements across vast expanses of the cosmos. They're what allowed astronomers in the late 1990s to detect the accelerated expansion of the universe, and they're what astronomers now use as their best handle on the nature of dark energy.

(It might sound odd to use massive stellar explosions as distance benchmarks, because, of course, we can't predict exactly when or where one will go off. But it turns out that the stellar explosion rate is high enough—a good rule of thumb is one supernova per galaxy per century—and there are so many galaxies, that if we just take pictures of lots of galaxies every night, we're likely pretty often to see a blip in one that wasn't there the night before, and then we can follow it up with more detailed observations.)

The precision with which we can now calibrate galaxy distances with supernovae is impressive, with accuracies pushing toward the 1 percent level. This makes it possible to measure the expansion rate of the universe, by determining how distant the galaxies are and how fast they're moving away. As discussed in Chapter 3, we talk about the expansion rate in terms of the Hubble Constant—the number that relates distance and recession speed. As of this writing, supernova measurements allow us to measure the Hubble Constant to an accuracy of 2.4 percent.

---

* This would be an excellent pun if Type Ia were likely to be able to create gold. While they can create other elements during the explosion (impressive amounts of nickel, for instance), due to the extreme temperatures and pressures involved, gold is probably mostly made in collisions of neutron stars. Alas.

Which is weird, because the number we get totally disagrees with the value of the exact same number we derive from looking at the cosmic microwave background.

EXPANDING CONFUSION

For the last several years, measurements of the Hubble Constant from supernovae have been giving us a number around 74 km/s/Mpc—that means that a galaxy one megaparsec away (that's around 3.2 million light-years) is receding from us at around 74 km/s. One twice as far away is moving, relative to us, about twice as fast. But we can also measure the Hubble Constant indirectly, by carefully studying the geometry of the hot and cold spots in the cosmic microwave background. When we measure it that way, the number we get is closer to 67 km/s/Mpc. Even though these observations are looking at very different epochs of cosmic history, each of them can tell us the expansion rate today. In a universe made of what we think it's made of, both methods of determining the Hubble Constant really ought to give us the same number. And they don't.

This hasn't always been considered to be *that* big a problem, since no one thought either measurement was so incredibly precise as to settle the question. Until recently, the state of play was that the cosmic microwave background folk assumed that there was some distance ladder mis-estimate that would be sorted out eventually, dropping the number down a tad, and the supernova folk figured that the CMB measurements, which derive ultimately from attempting to measure the shape of space itself, were so complicated that surely something would show that the number was really just a little bit higher. This isn't an unreasonable assumption, given the number of calculations and conversions that go into looking at a baby picture of the universe and converting that into a present-day expansion rate. And the distance ladder, likewise, really is fantastically complicated. Before even getting into all the possible biases that might creep in if you don't account for every relevant property of the supernovae themselves, calibrating variable stars is not easy, and even distances to relatively nearby galaxies sometimes come with huge uncertainties. Part of this is due to how the populations of Cepheid variables we can see nearby are different from those far away, and . . . well, I could go on. Let me just say there are *debates*.

While the assumptions from each side that the other has done something wrong haven't quite gone away, the situation is getting increasingly uncomfortable due to the fact that both sides are improving their methods, knocking out all known sources of measurement bias, and still finding numbers that ever more precisely do not agree with each other.

It's unclear what the solution to this problem will end up being. Maybe it really does come down to systematic errors in the data, or some problem with the measurements themselves. Maybe it's just a statistical fluke, as unlikely as that looks on the surface. Some of the most intriguing explanations involve dark energy that is not your garden-variety cosmological constant, but is instead something rather more ominous—something that could perhaps lead to a Big Rip. There's one hypothesis that would go a reasonable way toward fixing the discrepancy between the measurements: dark energy getting more powerful over time, in just the way you might expect from the early stages of a phantom-dark-energy-dominated cosmos.

We probably shouldn't panic just yet. As discussed, the data still aren't that clear. Most measurements of w give a value that is fully consistent with -1, and though it's true that values less than -1 are sometimes very slightly preferred, that preference isn't really statistically meaningful. As for the Hubble Constant disagreement, even if all the measurements are correct, nonapocalyptic explanations for the discrepancy—involving weird models of dark matter, or altered conditions in the early universe—are very much in the running. In fact, even tweaking dark energy wouldn't be enough to totally solve the problem, so it's not unreasonable to assume that the solution might lie elsewhere. And even if there *has* been a sharp upturn in the effects of dark energy in recent cosmic history, suggesting something like phantom dark energy, we still have a LOT of time before a Big Rip could possibly occur.

In fact, the one thing that all the universe-ending scenarios we've already discussed have in common is that they definitely aren't coming around anytime soon. As far as we can tell from our best understanding of physics, we have at least tens of billions of years before even the most extreme version of a sudden Big Crunch reversal could occur, and no Big Rip could be less than a hundred billion years off. A Heat Death, considered by most to be even more likely, would be so far into the cosmic depths of the future that we hardly have terms to describe it.

There is one possibility, though, that is decidedly more menacing than all the rest. It presents the prospect of a doomsday brought upon us by, in essence, a manufacturer's defect in the fabric of the cosmos itself. It's plausible, well described, and supported by the very latest results from the most precise fundamental physics experiments ever performed. And it could happen literally at any moment.

CHAPTER 6:

Vacuum Decay

None of the things one frets about ever happen. Something one's never thought of does.

Connie Willis, *Doomsday Book*

In March 2008, a retired nuclear safety officer named Walter Wagner filed a lawsuit against the U.S. government to prevent scientists from starting up the Large Hadron Collider. From Wagner's point of view, it was a desperate bid to save the world. The lawsuit was, of course, doomed to fail. For one thing, the LHC is controlled by the European Organization for Nuclear Research (abbreviated as CERN; the acronym comes from the French), not the U.S. government. And Wagner's scientific worries, though presumably sincerely felt, were unfounded. In the end, CERN leadership put out some reassuring press releases about the safety of its collider technology, and the construction and operation of the LHC continued.

That didn't stop some segments of the public from ramping up their panic as the date of the first scheduled particle collisions approached. The LHC would be the most powerful particle physics experiment in history, colliding protons in four places along a giant circular, supercooled, vacuum-sealed underground track 27 kilometers in circumference. These collisions would produce, within the detectors, momentary bursts of energy so powerful they could re-create the conditions of the Hot Big Bang mere nanoseconds after the first moment of creation. The LHC would, the scientists hoped, lend us insight not only into the conditions of the early universe, but into the very structure of matter and energy itself. Earlier experiments had shown us that the laws of physics are energy-dependent—altering how particles and forces interact depending on the conditions in which they're found—so creating collisions of higher and higher energies would allow scientists to probe the edges of our understanding of how physics works.

And there was an even more tantalizing prize in sight. Decades before, physicists had theorized the existence of a new particle—one so central to the behavior of matter that it would be the final piece completing the Standard Model of particle physics. The Higgs boson, if it were discovered, would, for reasons we'll get into shortly, finally confirm the leading theory explaining how fundamental particles were able to acquire mass in the early universe. And it would hopefully give us clues to the structure of physical law in regions beyond our current realm of exploration.

But that very prospect—probing the unknown reaches of reality—was enough to strike fear into the hearts of onlookers. No one had ever created collisions at these energies. No one knew how the laws of physics might shift and re-form themselves in such an environment.

Worst-case scenarios swirled around the internet. Perhaps the machine would open some kind of portal to another dimension, tearing apart the fabric of space itself. Perhaps it would create a tiny black hole that would grow, engulfing the entire planet. Perhaps it would create "strange matter"—a kind of composite material made of up-, down-, and strange-flavored quarks* that, some supposed, could lead to an ice-nine-style chain reaction,† converting all the matter it touched. But the physicists pressed on, apparently unconcerned. The LHC produced its first high-energy proton collisions in November 2009.

Given that life on this planet still exists, it's not too much of a spoiler to point out that none of the hypothesized existential disasters happened. (If you're still worried, there is a live-update website: www.hasthelargehadroncolliderdestroyedtheworldyet .com.) But did we just get lucky? Was the experiment really warranted, given the potential risks?

Physicists are not always cautious people, but exploring "what if" scenarios is kind of our bread and butter, and a chance to think deeply about the real physics behind hypothetical possibilities for ultimate destruction is too hard to pass up.‡ In fact, in 2000, four physicists (including one who would later win a Nobel Prize) wrote a sixteen-page paper for *Reviews of Modern Physics* called "Review of Speculative 'Disaster Scenarios' at RHIC." RHIC was the Relativistic Heavy Ion Collider, a Brookhaven National Lab collider that predated the LHC and that was built to collide the nuclei of heavy elements such as gold at high energies. A pioneering experiment in its own right, it was also the subject of worries that it might create unforeseen consequences that could endanger the planet (or the universe), and the paper was written to fully explore, and hopefully dispel, those rumors.

---

* Quarks come in six different "flavors," which have different masses and charges. The flavors are: up, down, top, bottom, charm, and strange. They were named in the 1960s.

† In Kurt Vonnegut's *Cat's Cradle*, a new form of ice is created, "ice-nine," that is more stable than liquid water. In the story, every bit of water a particle of ice-nine touches turns to ice-nine, creating an existential threat to life and the world.

‡ Believe me, I know.

The results were encouraging. Not only did the researchers find that the chances of producing strange matter or black holes were incredibly small based just on theoretical considerations, there actually was experimental data to back this up. Specifically: the existence of the Moon.

The argument for any kind of weird collider-induced phenomenon destroying us rests on the notion that the extreme high-energy collisions in these colliders are so unprecedented that we couldn't possibly know what might happen. Which is ignoring an important fact: while the energies reached by RHIC and LHC might be novel to us puny humans, cosmic rays cruising through the universe reach incredibly high energies all the time, and collide with other objects and each other constantly. In the words of the RHIC paper authors, "It is clear that cosmic rays have been carrying out RHIC-like 'experiments' throughout the universe since time out of mind." Collisions at much higher energies than what any Earthly collider could reach have been occurring across the universe for billions of years, so if they could destroy the cosmos, surely we would have noticed.

"Hang on," you might say, "what if cosmic ray collisions in deep space really are incredibly destructive, but just too far away to affect us? What if strange matter clumps exist all over the cosmos, and we just don't know?" It's a valid concern. While most of the time, particles produced in a collider are expected to have enough leftover momentum to zip out of the lab as soon as they're formed, it's conceivable that we could create something dangerous that could more or less come to rest in the detector. What then?

Fortunately, we can use the Moon as our canary in the coal mine. We have enough data from Earth-based detectors and space telescopes to know that high-energy cosmic rays slam into the Moon *all the time*. (In fact, with radio telescopes we can even use the Moon as a neutrino detector,* which is kind of amazing in and of itself.) If high-energy particle collisions could convert nearby ordinary matter into strange matter, this would have happened on the Moon eons ago, and we would have a VERY different object in our sky. Likewise, the night sky would be pretty noticeably altered if a tiny black hole formed on the Moon and swallowed it up. Not to mention the fact that we humans have actually *been there*, walked around, hit a few golf balls, and brought back samples. The Moon is doing just fine. Ergo, the authors argued, the RHIC won't kill us all.

Strange matter and black holes weren't the only apocalypses debunked, though. Another possibility, similarly dismissed by witnessing the superior firepower of cosmic rays, is the notion that a powerful enough collision could trigger a universe-destroying quantum event called *vacuum decay*. The whole idea of vacuum decay rests on the hypothesis that our universe has a kind of fatal instability built into it. While that might sound scary even as just as a remote possibility, at the time the RHIC was commissioned, there was no real evidence for such a flaw, so it wasn't taken especially seriously.

When the LHC discovered the Higgs boson, in 2012, that all changed.

THE STATE OF THE UNIVERSE

A good way to make a particle physicist cringe is to refer to the Higgs boson by the name that made it famous: the *God*

---

* This is due to something called the Askaryan Effect, in which an ultra-high-energy neutrino punches through the lunar regolith and creates a burst of radio waves that we can hopefully pick up with radio telescopes. So far, our telescopes haven't been sensitive enough, but we should be able to pick up these signals with the next generation of instruments.

Particle. Our collective grumpiness around this lofty moniker isn't fueled entirely by a discomfort with the mixing of science and religion (though for many that's a big part of it). It's also that "God Particle" is just terribly imprecise, and sounds a bit presumptuous, frankly. Which is not to say that the Higgs boson isn't a deeply important part of the Standard Model of particle physics. It could even be argued that the Higgs is key to everything else fitting together. But it's really the Higgs *field*, not the particle, that plays a central role in the workings of particle physics and the nature of the cosmos.

The short version of the story is that the Higgs is a kind of energy field that pervades all of space and has interactions with other particles in a way that allows them to have mass. The Higgs *boson* has the same relationship to the Higgs field that the photon, the carrier of the electromagnetic force (and light), has to the electromagnetic field—it's a localized "excitation" of something that pervades a larger space. The long version of the story has to do with electroweak theory, the theory that unites the weak nuclear force with electricity and magnetism, and how a process called "spontaneous symmetry breaking" separates those forces.

(This is the part of the book where I really really want to teach you all of quantum field theory, but where by a heroic effort I limit myself to just touching on a few key issues. You'll just have to trust me that if you decide to go and learn the mathematics behind all this, it gets MUCH cooler.)

We talked in Chapter 2 about the fact that physics works differently at different energies. Electromagnetism and the weak nuclear force, for instance, act like fully separate phenomena at the kinds of energies we deal with in everyday life, but in the very early universe, at very high energies, they were aspects of the same thing. The Higgs field was instrumental in that transition; when it changed, the laws of physics changed too.

This is a big part of why we build colliders: to create, in tiny little spaces inside our detectors, the kinds of extreme conditions that existed at the beginning of the universe and that can give us insights into the underlying physical principles that dictate how *everything* in physics fits together. The basic idea is that there must be some kind of overarching mathematical theory that gives us a blueprint for particle interactions under all possible conditions, and by continually producing higher and higher energy interactions, we get an increasingly clear picture of what that larger framework looks like.

As an analogy, think of water. At the most fundamental level, it's a collection of molecules made up of bonded hydrogen and oxygen atoms in a particular arrangement. But our everyday experience of water is as a uniform colorless liquid, or perhaps as a crystalline solid, or at certain unfortunate times as the kind of soul-crushing humidity that makes you wish your clothing were made of towels.* By examining water's behavior in these different forms, we can make inferences about what it *really is*, even if we don't have powerful microscopes at hand to see the individual atoms themselves. The shape of a snowflake, for example, tells us something about the shape of the molecules as they arrange into crystals. The way water evaporates tells us something about the bonds that hold the molecules together. If we only ever experienced water in one of its phases, we wouldn't have a full picture of it, and it would be harder to get to that complete story. In the same way, our experience of subatomic particle interactions changes based on the energy (or temperature) of the experiment, and that allows us to get a better view of what's really going on.

What we want to know in particle physics is how the particles interact with each other and how their fundamental properties, like their masses, came to be what they are. The salient feature of any particle that has mass is that it can't accelerate

---

* This part of the book was written in North Carolina in August.

135

without an application of force, and it can never reach the speed of light. In the very early universe, the Higgs field underwent a transition that separated the electroweak force into electromagnetism and the weak force, and in the process gave some particles (though not the photon or gluon) the ability to interact with the Higgs field itself. The strength of that interaction determines the particle's mass. The photon continues to zip through space at the speed of light, but the particles with mass move more slowly in proportion to the tugging they experience from the Higgs.

Comparing the behavior of particles in the early universe to their behavior today is like comparing how you might interact with vapor versus liquid water. Imagine the vapor is the Higgs field—an energy field, present at every point in space. And imagine that at some point that Higgs field drastically changes character, as completely as vapor condensing into liquid water. If you've been used to encountering nothing but humid air, moving through a pool of water is a completely different prospect. When the Higgs field suddenly shifted in character, it was as though the laws of physics had condensed into a totally different form. Suddenly, particles that could move through space unimpeded at the speed of light were slowed down by their interactions with the Higgs field. They obtained mass.

We call this process *electroweak symmetry breaking*.

FEARFUL SYMMETRY

Symmetry in physics is the kind of subtle, abstract concept that is extremely hard to explain without equations but which is also so absolutely vital to everything we think about as physicists that I can't in good conscience just brush by it. Symmetry is central to how we describe theories of nature, and, more often than not, to how we develop new ones. If you happen to be someone who is used to thinking about the world in terms of the mathematical equations that govern it, you're probably already comfortable with the idea that theories can be described in terms of the symmetries they obey; if you're not, that's complete gibberish to you, and understandably so. So let's take a short detour for a moment and lay this all out, because it's an incredibly beautiful thing, and once you know about it you see it everywhere.

Symmetry isn't just about whether or not something looks the same in a mirror. In physics, it's all about patterns, and how those patterns can give you deeper insight into some underlying structure. Take, for instance, the periodic table of the elements. Why are the elements arranged in the rows and columns we are used to seeing them in today? If you've studied chemistry, you'll know that certain columns collect elements that have things in common—the noble gases, the column on the far right, are all loath to chemically react, whereas the halogens, right next to them, are especially volatile. These patterns were discovered before the table was even complete, and in fact the creator of the table, Dmitri Mendeleev, left gaps for elements he knew *should* exist based on the pattern, even before they were discovered.

The patterns in the periodic table led to theorizing about electron orbitals, which led to discoveries about the fundamental nature of subatomic matter. Over and over again, scientists have developed new theories of nature by recognizing patterns in their observations and then looking for a hidden property that could give them insight into what was really going on. We do this all the time, ourselves, without noticing it. Watching highway traffic change over the course of the day can tell you standard business hours. The pattern of fading in a carpet can let you deduce which parts of the room get the most sunlight (and thereby, indirectly, tell you how the Earth and Sun are oriented in the Solar System).

In the case of particle physics, using symmetry is often a lot like building new periodic tables, but for even smaller building blocks of nature. Similarities between particles—their charge, mass, or spin, for instance—can give clues about similarities in their formation or their connections to fundamental forces. Arranging these particles by their patterns lets physicists identify the symmetries that can be the defining features of entire theories.

Sometimes those patterns are most easily seen mathematically. If you write down an equation to describe a physical process, and then find that you can swap some terms around without actually changing the physical phenomenon the equation describes, you've found a mathematical symmetry. And it's probably telling you something deep about the particles or fields you're describing.

This symmetry-oriented way of looking at particles and the relationships between them is so prevalent in physics that we find ourselves using references to mathematical symmetries as shorthand for the theories themselves. For instance, electromagnetism is frequently called U(1) theory, because some aspects of the mathematics have the same kind of symmetry as a circle, and "U(1)" is shorthand for a mathematical group that describes rotations around a circle.

A symmetry *breaking* event is when the conditions suddenly change such that the theory you would write down to describe how particles interact takes on a different, less symmetric, structure. After a symmetry breaking occurs, you can no longer swap around symbols in equations in the same way, and this change in symmetry expresses itself as altered behavior in the physical world.

Some of the symmetries we work with in physics are abstract and only obvious in the mathematics, but some are the usual stuff. Rotational symmetry is when something looks the same rotated by some angle (like a circle or a five-pointed star). Translational symmetry means something looks the same if you shift it to one side (e.g., a long picket fence moved over the distance of one picket, or a long straight line slid over by an inch). Breaking a symmetry involves doing something to the situation to make that symmetry no longer work. A wineglass has perfect rotational symmetry until a lipstick mark appears in one spot. A picket fence has translational symmetry until one of the slats is broken. Even a dinner party can include a symmetry breaking event, as is frequently observed by groups of physicists at conference banquets after the liquor comes out. As you wait patiently at the start of the meal, surrounded by a perplexing array of silverware, a small bread plate on either side of you, you are in a rotationally symmetric situation. As soon as one person reaches right or left to take up the bread plate, the symmetry is broken, and everyone else can follow their lead.*

No matter what kind of symmetry we're working with, as physicists, we'll see it in the equations describing the inter-actions. There are ways to encode rotational, reflection, and translational symmetries in equations, so you know the phys-ics stays the same no matter how you rotate, flip, or move the system in question. Equations can also encode subtler kinds of symmetries best described using group theory and abstract algebra that are FASCINATING but sadly pretty far outside the scope of this book.

When electroweak symmetry breaking occurred, way back when the universe reached the ripe old age of 0.1 nanosecond, it was a kind of rearranging of the structure of physics on a

---

* Two people reaching at the same time for bread plates on opposite sides of them results in a pile-up that physicists would call a topological defect. In this specific case, it would be a domain wall, which, if let loose on the cos-mos, would dominate the universe and lead to a Big Crunch. This is why I always wait for someone else to select the bread before making my attempt.

fundamental level.* The rules particle interactions must follow are completely different in our post-electroweak-era universe. The previously vaporous Higgs field has become an ocean.

The water analogy isn't perfect. When you move through water, you're slowed down by drag, which means that you'll come to a halt if you stop expending effort. In the case of massive particles interacting with the Higgs field, the interactions don't slow them down over time. Anything moving through a vacuum tends to keep doing what it's doing. In the case of massive particles, this frequently includes careening through the universe at very high (though sub-light) speeds. The main difference between massive and massless particles is that, in order to change speed, massive particles moving through a vacuum require a push, whereas massless ones travel at light speed effortlessly. In fact, massless particles can't travel at anything but light speed.

So we, who enjoy the ability to sit still once in a while, should be grateful the Higgs field did what it did, and broke the electroweak symmetry. The Higgs field not only gives particles the ability to have mass, it also determines several of the fundamental constants of nature, like the charge of the electron, or the masses of particles. The particular physical state we live in, with the Higgs field nicely situated where it is, is referred to as our "Higgs vacuum" or "vacuum state." If the Higgs field had some other value, or if the symmetry had broken in some other way, we might not be able to exist at all. We enjoy a universe in which the masses and charges of particles are perfectly set to allow them to come together in molecules, form structures, and carry out the chemical processes of life. If the field took some other value, this delicate balance might be off, potentially making these bonds impossible. We owe our entire corporeal existence to the fact that the Higgs has settled on the value it has.

---

* We previously discussed this transition, and what it meant for the very early universe, in Chapter 2.

And this is where things start to get a bit dicey.

Experiments like the LHC, which create extreme conditions that mimic those of the early universe, help us to see not just what the laws of physics are, but what they could be, under different circumstances. In 2012, when physicists finally were able to produce the Higgs boson in particle collisions, measuring its mass produced the final missing puzzle piece in the Standard Model of particle physics. It gave us a glimpse not only of the current value of the Higgs field, but of all the possible values it might take, given half a chance.

The good news is that the measurement of the Higgs mass is in perfect agreement with a nicely reasonable and mathematically consistent formulation of the Standard Model that has so far passed every experimental test with flying colors.

The bad news is that this consistent picture of the Standard Model also tells us that our Higgs vacuum—the perfectly balanced set of laws that govern the physical world—is not stable.

Our whole beautiful cosmos appears to be living on borrowed time.

A SLIPPERY-SLOPE COSMOS

The idea that our vacuum might not be stable isn't a new one. Even in the 1960s and 1970s, physicists were gleefully writing papers imagining the possibilities for a universe that might undergo a catastrophic decay process, destroying all life as we know it and even all possibility of organized matter. Of course, at the time, vacuum decay was just a fun idea to play around with in the equations, with no experimental data to back it up.

Unlike now.

To understand vacuum decay, we have to understand the concept of a *potential*, a mathematical construct that represents how the value of a field can change, and where it "prefers" to be. You can think of the Higgs field as a pebble rolling down a slope into a valley, with the potential represented by the shape of that slope. Just as a pebble will settle in the bottom of a valley, the Higgs field will seek the lowest-energy state, where the potential is at its lowest value, and settle there, assuming nothing stops it. A sketch of the potential might look like a U shape, with the bottom of the U being the bottom of that valley. When electroweak symmetry breaking happened, it created the potential that the Higgs field is governed by, and as we generally imagine it, the Higgs is now safely settled at the bottom.

The problem is, that might not really be the bottom. There might be another vacuum state, at some even lower part of the potential. Imagine a kind of tilted rounded W shape, with one of the valleys, the one representing where our Higgs field doesn't live, a bit lower than the other. If the Higgs potential has that second, lower valley, it suddenly goes from being a nice mathematical construct to an existential threat to the cosmos.

Wherever the Higgs field is now in its potential, it's given us a perfectly livable, comfortable universe. We have constants of nature that are nicely compatible with bound particles and solid, life-compatible structures. If another state is possible, lower down the potential, all that is at risk.

In such a situation, the Higgs vacuum is only *meta*stable. Stable . . . ish . . . for now. The field is stuck in a part of the potential that looks like the bottom of a valley but is actually more like a divot in the valley wall. It can stay tucked in there for a long time—plenty long enough for the growth of galaxies, the birth of stars, the evolution of life, and the production and distribution of more superhero movies than anyone could ever really want—but the possibility looms that a large enough disturbance could kick it over the edge, and then there would be nothing to stop it from landing on the *actual* valley floor. And that would be really, really, apocalyptically bad. For reasons we will shortly discuss in gory detail.

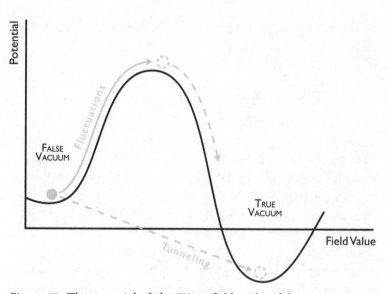

Figure 17: The potential of the Higgs field with a false vacuum state. Each valley in the potential is a possible state of the universe. If our Higgs field lives in the higher valley (the false vacuum), it could transition to the other state (the true vacuum) via a high-energy event (marked "fluctuations" on the diagram) or via quantum tunneling. If we live in a false vacuum universe, a transition of the Higgs field to the true vacuum would be catastrophic.

Unfortunately, the best data we have, consistent with every measurement of the Standard Model of particle physics, suggests that our Higgs field is currently clinging onto just such a divot. This metastable state is also known as a "false vacuum," as opposed to the "true" vacuum at the bottom of the valley floor.

What's wrong with being in a false vacuum? Quite possibly, everything. A false vacuum is at best a temporary reprieve from ultimate destruction. In a false vacuum, the laws of physics, including the ability of particles to exist at all, are contingent on a precarious balancing act that could be upset at any moment.

When this happens, it's called vacuum decay. It's quick, clean, painless, and capable of destroying absolutely everything.

A BUBBLE OF QUANTUM DEATH

In order for vacuum decay to occur, there has to be a trigger—something that will set the Higgs field wandering far enough to find the part of the potential corresponding to the "true" vacuum and realize it would rather be there.* An ultra-high-energy explosion, or the catastrophic final evaporation of a black hole, or even an unfortunate quantum tunneling event (more on these later) could set it off. If this happens anywhere in the cosmos, it creates an unstoppable apocalyptic cascade that nothing in the universe can withstand.

It starts with a bubble.

Wherever the event occurs, a tiny bubble of true vacuum forms. This bubble contains within it a drastically different kind of space—one in which the processes of physics follow different laws, and the particles of nature are rearranged. At the moment it forms, it's an infinitesimal speck. But it is already surrounded by a bubble wall of extremely high energy that could incinerate anything it touches.

Then, the bubble begins to expand.

Because the true vacuum is the more stable state, the universe "prefers" it, and will revert to it if given the slightest chance, just like a pebble will roll down a slope if it's placed on one. As soon as the bubble appears, the Higgs field all around it is suddenly being shaken down to the valley floor. It's as

---

* Of course, the Higgs field doesn't have preferences; it's just governed by its potential. But the way it would dive into a true vacuum if it could would definitely give an impression of enthusiasm.

though that first event knocks free every precariously bal-anced pebble near it, and then the avalanche spreads. More and more space succumbs to the true vacuum state. Anything unfortunate enough to be in the bubble's path is first hit by the intensely energetic bubble wall, approaching at about the speed of light. Then it undergoes a process that could only be called total and complete dissociation, as the forces that previ-ously held particles together in atoms and nuclei can no longer function.

Maybe it's for the best that you don't see it coming.

As dramatic as the process sounds from a bird's-eye view, if you happen to be standing nearby when the bubble appears, you won't notice it. Something coming at you at the speed of light is invisible—any little glint warning you of its approach arrives at the same time as the thing itself. There is no possible way to see it coming, or even to know that anything has gone wrong. If it approaches you from below, there will be a couple of nanoseconds during which your feet no longer exist while your brain still thinks it is looking at them. Fortunately, the process is also entirely painless: at no point will your nerve impulses be able to catch up with your disintegration by the bubble. It's a mercy, really.

Of course, the bubble doesn't stop with you. Any planet or star within its ever-expanding radius suffers the same fate, equally oblivious to what's coming. Entire galaxies are engulfed and obliterated. The true vacuum cancels the universe entirely. The only regions able to escape are those that lie so far away that the accelerated expansion of the universe keeps them beyond the bubble's horizon forever.

In fact, it's entirely possible that, as we sit here now, calmly drinking our tea, vacuum decay has already occurred. Maybe we're lucky and the bubble is beyond our cosmic horizon, swallowing up galaxies we would never have known. Or maybe it is, cosmically speaking, right next door, quietly approaching

Figure 18: The bubble of true vacuum. If a vacuum decay event happens at one place in the cosmos, it causes a bubble to expand outward at the speed of light, destroying everything in its path.

with relativistic stealth, destined to catch us unawares, between breaths.

KICKING THE HORNET'S NEST

You shouldn't worry about vacuum decay. Really. For several reasons. There are the obvious ones, of course: there's no way to stop it if it's happening; and you can't know it's about to; and it's not like it would hurt; and no one would be around to miss you anyway, so what's the point of worrying about it? You're better off double-checking your smoke alarm batteries and, I don't know, lobbying to close down coal power plants or something. But if for some reason that isn't sufficiently reassuring, I can also say with a reasonable degree of certainty that vacuum decay is extremely unlikely to happen—at least, anytime in the next many many many trillions of years.

There are a few ways vacuum decay could, in theory, occur.

The most straightforward is some kind of high-energy event. Think of it as the equivalent of an earthquake, knocking the pebble out of its divot to send it plummeting to the valley floor. Fortunately, the "earthquake" in this case would have to be really unfathomably powerful. The best estimates we have suggest that the event would have to be much more energetic than the most devastating explosions we've witnessed in the cosmos, and certainly many orders of magnitude stronger than anything we could possibly do with a human-built machine like the Large Hadron Collider. If we're ever worried about that, we can always appeal again to the fact that particle collisions in the cosmos are and have always been reaching much higher energies than the LHC or any other machine possibly could, so as long as we haven't blinked out of existence yet, our modern equivalent of banging rocks together is really no threat at all.

The difficulty of creating a high enough energy event to directly trigger vacuum decay comes down to the height of the *potential barrier* between our false vacuum and the true one. Going back to the picture of the pebble stuck in a divot, the potential barrier is the bit of land that sticks up to make a divot pocket-shaped. In our current best guess at the true shape of the Higgs potential, the divot is a substantial one, separated from the deeper true-vacuum valley by a very high ridge. The amount of energy it would take to kick the pebble over that ridge (or push the Higgs field over its potential barrier) is so high, it's hardly worth worrying about.

Except . . . we're living in a universe that doesn't follow those kinds of rules. Our cosmos is fundamentally based on quantum mechanics, and in quantum mechanics, if you're living on a subatomic scale, the path you take to get from one place to another might, very rarely, send you sailing right through solid objects without missing a beat. If you're standing in front of a wall, you might not need to get enough energy to jump over it.

You might be able to step right through it instead. Especially if "you" are the Higgs field.

TUNNELING INTO THE ABYSS

Quantum tunneling sounds like science fiction or some obscure theoretical thing that physicists sit around chuckling over while writing down incomprehensible equations. Like, sure, quantum mechanics says that you can't ever really say for sure exactly where a particle is, or what path it's taking as it travels. That means that to get the math to work out, you have to write down and calculate things about *all* the paths, even the outlandish ones that send the particle from one side of the lab to the other by way of a coffee shop three cities over. But that doesn't mean that the particle *really* does that, right?

As it happens, the question of what the particle *really* does is surprisingly hard to answer, and has spurred a decades-long debate about interpretations of quantum mechanics. Where the particle goes on the journey between Point A and B is still something of a mystery, as is what it actually *means* that particles are *measured* as small localized things but still manage to obey the *mathematics* of waves that are spread out through space.

The one thing everyone agrees on is what the data say, and those data make it very clear that tunneling through seemingly impassable barriers is something that particles are very happy to do on the regular. Wherever a particle really goes in the interim, it's clear that a wall can't stop it. This sort of escape artistry is such normal behavior for particles that people who design things like cell phones and microprocessors have to take into account the fact that every once in a while an otherwise well-behaved electron will suddenly materialize on the wrong side of a chip. Some technologies, including flash memory, occasionally also use this to their advantage. Scanning tunneling microscopes use the expectation of tunneling almost like a valve to drip electrons slowly onto a surface and get images of individual atoms.

Letting electrons sneak across short gaps or squeeze through insulating barriers is a nice party trick, but it gets significantly more ominous when you realize that quantum tunneling can be performed not only by particles, but also by fields. Fields like the Higgs, separated from that big true-vacuum valley by a potential barrier it can tunnel right through. The only thing standing between our nice hospitable universe and ultimate cosmic disaster suddenly looks a lot less solid.

The (somewhat) good news is that even something as weird as quantum tunneling does actually follow certain rules, at least when it comes to the expected rate of its occurrence. The probability for a tunneling event is based on the physical characteristics of the system, which means that how likely it is to happen over a set period of time can be very well known. It isn't a total free-for-all. As hard as quantum mechanics may be to fully understand or interpret, it is at least calculable.

But those "rules" that we calculate don't come in any form more reassuring than probabilities. We can't confidently say that the Higgs field *won't* tunnel across the barrier and create a bubble of quantum death right next to you in the next 30 seconds, setting off a process of unthinkable destruction that will tear through space for all eternity. What we can say is: that scenario is extremely unlikely. (At least, the "in the next 30 seconds" part is. If our vacuum really is metastable, strictly speaking, the bubble has to show up eventually.)

The best calculations we have suggest that our nice pleasant vacuum is not likely to undergo a radical rearrangement anytime soon—at the time of writing, the latest estimate gives us more than 10^{100} years. By then, we'll likely be well into the process of a Heat Death, or perhaps, if we're very unlucky, being torn asunder by a Big Rip. At which point maybe an instant painless obliteration won't seem so bad.

So, technically, I can't tell you for certain that vacuum decay isn't right about to happen. I also can't tell you for certain that it hasn't already happened somewhere in our own Solar System, or on the other side of the galaxy, or in another galaxy, creating a light-speed-expanding bubble that is silently coming for us as we speak. But I can tell you that if you'd like to prioritize your paranoia, you're *much* more likely in your lifetime to be struck by lightning, an out-of-control car, rampaging cattle, or even a stray meteor than by the spontaneous appearance of a bubble of true vacuum.

There's just one more thing, though.

We've already covered the facts that we can't produce our own vacuum decay bubble with a high-energy particle collision, and that a spontaneous tunneling event is so unlikely we should probably try very hard to forget that we ever heard of it in the first place. But recently, physicists have come up with yet another way to destroy the universe with vacuum decay, and I have to say it's kind of a cool one.

SMALL BUT DEADLY

In 2014, Ruth Gregory, Ian Moss, and Benjamin Withers, building on a bit of previous work on this topic, put out a new paper that caught my attention. It explained that while spontaneous vacuum decay is tediously slow, the presence of a black hole could speed the process up considerably and generally make things much more interesting. In fact, they argued, the really dangerous thing is a *small* black hole, because particle-sized black holes can dramatically increase the chance of vacuum decay occurring right on top of them. Maybe we don't have to wait 10^{100} years after all.

The way it works is similar to the way a particle of dust can condense a bit of water around it in a humid room, or the way that clouds get seeded in the upper atmosphere. The dust particle is a *nucleation site*—something that differentiates that point from others and allows the process to happen more easily. In the case of clouds and water, the water molecules have an easier time sticking together if there's something else that they can stick to first. So an impurity can set off a chain reaction where otherwise things might have continued on as they were. It turns out tiny black holes can be those nucleation sites for bubbles of true vacuum, but only if they're very small.

Fortunately for the universe, tiny black holes are not easy to make, given our current understanding of gravitational physics. Generally speaking, we only expect black holes to form at masses greater than that of the Sun, as the result of the collapse of massive stars at the end of their lives. Those black holes might grow to much larger masses by pulling in material or by merging with each other, but shrinking is another matter entirely. They can only lose mass via Hawking evaporation (see Chapter 4), and that takes ages. A black hole as massive as the Sun has an expected lifetime somewhere in the vicinity of 10^{64} years. At some point toward the end of that time, the black hole might get small enough to trigger vacuum decay, but we have quite a while before we really have to worry about that. It's also been hypothesized that in the early universe, tiny black holes might have formed due to the extreme densities of the Hot Big Bang, but so far we don't have any evidence of those. If they did form, though, and if little black holes really could destabilize the vacuum, we wouldn't be here. So if we take that into account, and believe in the possibility of vacuum decay, any theory that predicts tiny primordial black holes has to be wrong, because we exist.

Just for fun, a few of us have also been wondering whether there are ways to make those little black holes without them having been around since the very beginning of the universe. Making tiny black holes isn't a new idea. In addition to being awfully cute in a terrifying theoretical kind of way, these mini monsters could teach us about how gravity works, whether or not black holes really do that cool evaporation thing, and even whether or not there might be additional dimensions of space that we can't otherwise see.

For years, physicists have been combing the data from particle colliders, hoping to see some telltale sign that one of the collisions between protons managed to put so much energy into such a small space that it all immediately collapsed into a microscopic black hole. That black hole, if it appears, *should* be harmless, according to the traditional thinking, not taking the possibility of vacuum decay into account. Theory states that it should immediately evaporate via Hawking radiation, and that even if it didn't, it would likely be moving at relativistic speeds in some direction that would take it far away from us in a very short time, because a collision is never so perfectly timed and aimed that the particles completely stop. Plus, for the kinds of collisions that happen in particle colliders to be capable of making tiny black holes, somehow the force of gravity felt by subatomic particles has to be stronger than Einstein's laws of gravity would suggest. And the only way *that* could happen, as far as we know, is for there to be extra dimensions of space. We'll talk more about this in the next chapter, but the short story is that having more than our usual three dimensions of space can make gravity a bit stronger on very small scales and can therefore allow LHC collisions to make little black holes.

So if we can make black holes in the LHC, we have evidence for space having more dimensions than we thought. Which to a physicist looking for signs of exciting new physics seems like it would be fantastic news! Of course, it would be a shame if those little black holes we are trying to make in the LHC could trigger vacuum decay and cause the universe to end . . .

Fortunately, they can't. We are as close to absolute certainty about that as physicists ever get. The main thing that acquits them is the fact that, as previously stated, cosmic rays can make collisions much more powerful than anything we see in our own colliders. If we can smash protons together to make black holes, the universe has already done this countless times, and, look!—we're still here! So either the black holes aren't being made anywhere, or they were harmless all along.

The other reason is that it seems like there's a mass threshold tiny black holes have to reach before they're even hypothetically dangerous. The kinds of black holes a particle collider could make would be safely below that level, as would, most likely, many of the collisions that would happen out in space. As a side bonus, some of us have already been working on using this fact, along with our continued existence, as a way to argue that there have to be limits on the possible size extra dimensions could be.* (Just personally, as a cosmologist interested in testing different theories of physics, it's always fun to be able to hold up the lack of a cosmic apocalypse as a data point.)

So, setting aside the little black holes for now, where does that leave us with vacuum decay? All the other possible ends of the universe we've looked at offer us at least the small comfort of being so far in the future that we can, with a great deal of confidence, leave them to be worried over by whatever post-human entities might inhabit the cosmos after we're gone. Vacuum decay is special in that it could technically happen at any moment, even if the chances of that are astronomically low. It also comes with a uniquely extreme, almost gratuitous finality.

---

* "Some of us" here being, specifically, me and my colleague Robert McNees, in our 2018 paper in *Physical Review D*. It was a fun one.

In 1980, two theorists, Sidney Coleman and Frank De Luccia, calculated that a true vacuum bubble would contain not only a totally different (and lethal) arrangement of particle physics, but also a kind of space that is, by its nature, gravitationally unstable. Once the bubble formed, they explained, everything inside would collapse gravitationally within microseconds. Then they wrote:

> *This is disheartening. The possibility that we are living in a false vacuum has never been a cheering one to contemplate. Vacuum decay is the ultimate ecological catastrophe; in a new vacuum there are new constants of nature; after vacuum decay, not only is life as we know it impossible, so is chemistry as we know it. However, one could always draw stoic comfort from the possibility that perhaps in the course of time the new vacuum would sustain, if not life as we know it, at least some structures capable of knowing joy. This possibility has now been eliminated.* *

THE JOY OF NOT KNOWING

Of course, vacuum decay is, relatively speaking, a fairly new idea that incorporates so many kinds of extreme physics that it's entirely conceivable that our perspective on it will shift dramatically over the next few years. It may be that more detailed, rigorous calculations will give us different answers. These questions are difficult and complicated, and we still have a way to go before a consensus is reached.

If we conclude that our vacuum really is metastable, this may be incompatible with the theory of cosmic inflation. The

---

* This discussion remains, to me, one of the most beautiful pieces of physics poetry I've ever seen in an academic journal.

quantum fluctuations during inflation, or the ambient heat afterward, seem like they should have been sufficient to trigger vacuum decay in the first moments of the cosmos, negating our very existence. Clearly, that didn't happen. Which suggests either we don't understand the early universe, or vacuum decay was never possible at all.

Whether or not you trust early universe theories, taking vacuum decay seriously depends on placing a great deal of trust in the Standard Model of particle physics, which we know cannot be the whole story. Dark matter, dark energy, and the incompatibility of quantum mechanics and general relativity all point to there being something more to the universe than what we can currently write down. Whatever comes along to replace the Standard Model might, by the by, save us from even having to vaguely worry about a wayward bubble of quantum death.

Or it may be that extensions of fundamental physics present entirely new ways for the universe to end. The possibility of extra dimensions of space—the same ones that tantalize collider physicists hoping to make miniature black holes—extends the universe into new realms of the unknown. Like any explorer reaching the edges of the map, we reach out not knowing what we might find. Higher dimensions of space might allow us to solve some long-standing problems with our theories of gravity, but they also come with a warning, scrawled in the margins of the ever-growing cosmic map: here be monsters.

CHAPTER 7:

Bounce

HAMLET: O God, I could be bounded in a nutshell, and count myself a king of infinite space, were it not that I have bad dreams.

William Shakespeare, *Hamlet*

On September 14, 2015, at 9:50 a.m. and 45 seconds UTC, you were, for the briefest moment, just a little bit taller.

The gravitational wave crest that washed through you had been traveling across the cosmos, warping space itself in its wake, for 1.3 billion years, ever since it was set off by the violent merging of two black holes each 30 times more massive than the Sun. You might not have noticed the boost—after all, you grew by less than one millionth the width of a proton—but physicists at the Laser Interferometry Gravitational-Wave Observatory (LIGO) did. The first detection of gravitational waves was the culmination of a decades-long search, requiring the development of new technologies and the creation of the most sensitive equipment in the history of experimental physics. Finally detecting those ripples in spacetime was heralded as the ultimate vindication of Einstein's general theory of relativity.

But even more significantly, it was the dawning of a new age of astronomical observation. It opened up the universe to a totally new way of seeing. Instead of collecting light or high-energy particles from distant sources, we could now reach out and feel the vibration of space itself, creating for the first time a window onto the kind of distant cosmic violence that can shake the very foundations of reality.

Since that first discovery, gravitational wave astronomy has continued to show us the inspirals and catastrophic mergers of black holes and neutron stars and allowed us to study the workings of gravity with an unprecedented level of precision. But gravitational waves may hold the key to something even more fundamental. They may give us a new view of the shape and origin of our universe, and present us with the possibility of determining whether or not there might be something outside it. Something that may ultimately destroy it all.

THE UNBEARABLE WEAKNESS OF GRAVITY

We've known for a long time that something has to be wrong with gravity. It works too well. Einstein's general relativity has

Time

Figure 19: Illustration of the effect of a passing gravitational wave. As a gravitational wave hits face-on, it stretches the space it's moving through vertically while squeezing it horizontally, and then vice versa, with each wave crest. If you are in the path of the wave, you are alternately a bit taller and thinner, and a bit shorter and wider, over and over until the wave passes. The magnitude of the stretching of your body is only about one millionth the width of a proton.

so far performed perfectly in every situation in which it's been tested. For decades, physicists have tried to find some kind of deviation, somewhere, anywhere, that would show us how the simple* equations written down in Einstein's theory inevitably break down. Somewhere, in some extreme regime, like at the edge of a black hole or among the particles at the center of a neutron star, the equations must have some kind of a crack. We haven't found it in any of our searches so far, but we're sure it has to be there.

We have good reasons for being suspicious. Compared to the other forces, gravity is an oddball. Not only does it look totally different from a mathematical point of view, it's way too weak. Sure, when you get together enough mass for a galaxy, or a black hole, it seems fairly strong. But in daily life, it's easily the weakest force you encounter. Every time you lift a coffee cup you're overcoming the gravitational pull of the entire planet. It takes putting the mass of the Sun into something the size of a city before gravity can even begin to compete with the atomic and nuclear forces holding atoms together.

Comparing forces is about more than just a strength test, though. The idea that all the forces can somehow be reframed as different aspects of the same thing, in extremely high-energy environments, is generally considered to be key to truly understanding how physics works. We live in hope that there's some ultimate theory out there—a theory of everything—that unites all the forces of particle physics and gravity to explain, well, everything.

But so far, gravity won't play along. We have a rock-solid theory of the electroweak force (a unification of electromag-

---

* "Simple" here may be a matter of perspective. Working with the equations of general relativity requires a deep understanding of differential geometry, which is the sort of thing you pretty much only study if you're doing graduate work in physics or mathematics. But if you ARE such a person, the equations are as elegant and transparent as fine-blown glass.

netism and the weak nuclear force), confirmed by experiments. We also have some perfectly promising leads on a Grand Unified Theory uniting the electroweak and strong nuclear force. But every time we try to bring gravity in, its feebleness ruins the whole picture. Even aside from that, gravity and quantum mechanics (which describes the workings of all the other forces) explicitly clash in their predictions about things like what should happen at the edge of a black hole. Finding a way to bring gravity into line would help immensely.

So there seem to be a few options here. One obvious one is to abandon the whole idea of unification, and just let gravity be its own special snowflake of a theory, unconnected to the rest of physics. It's totally possible that there's no theory of everything, and we're never going to piece it all together in any sensible way. But just typing that makes my physicist toes curl, so maybe we can set that to the side for now in the "BREAK GLASS IN CASE OF EXISTENTIAL EMERGENCY" cabinet.

A much more appealing and intellectually exciting idea is that the problem is our theory of gravity: general relativity needs to be altered or replaced, and when that happens, it'll all fit together. There's been no shortage of impressive, well-motivated attempts in this direction. Quantum gravity theories, of which string theory and loop quantum gravity are the most famous examples, continue to be hot topics among theorists trying to find a way to unite particle physics with gravity and tie it all up with string. Or loops. You get the idea. In each of these scenarios, you end up with a gravity theory that can be *quantized*—expressed in terms of particles and fields rather than forces or spatial curvature—and these particles and fields mesh nicely with those of the quantum field theories that explain interactions between quarks and electrons and photons and the whole subatomic world. In this picture, gravitational forces would be manifestations of the exchange of particles called gravitons, just like an electric field is due to photons moving between objects. And gravitational waves, which we currently think of as the stretching and squeezing of spacetime, could also be envisioned as the motion of gravitons expressing their wavelike nature.

Unfortunately, despite decades of hard work and extraordinarily intricate calculations, we haven't yet settled on a theory that's broadly accepted by the physics community. Not only have none of the ideas put forward been confirmed by particle experiments, it's not even clear that they *can* be. Ideally, we'd write down two theories and then work out that they make different predictions for experiments like those being carried out at the Large Hadron Collider. But this is a challenge when you're trying to distinguish between theories whose effects only become clear at energies many orders of magnitude higher than what LHC collisions can produce. This has led physicists to suggest solutions ranging from abstract arguments aimed at narrowing down the total range of possible universes to philosophical debates about how to make advances in areas of theory in which experimental evidence may never appear.

For those of us who hold out more hope for new data, our best bet for something that could give us hints about a theory of everything might well lie in cosmology—especially the study of the early universe. If you need data about particle interactions at extraordinarily high energies, finding new ways to examine the Big Bang is generally going to be easier than trying to build a particle collider the size of the Solar System.

We're already being nudged in this direction. So far, we've seen only a small handful of physical phenomena that we cannot explain within the Standard Model of particle physics (or very minor modifications thereof). The big ones, dark matter and dark energy, are strongly supported by observational evidence. But absolutely all of that evidence comes from cosmology and astrophysics. Figuring out what these mysterious cosmic components are and how they work might be the best hope we have to see where the theories should go next.

Another thing pointing us toward cosmology is the strange imbalance between matter and antimatter in the universe. Where our current theories suggest matter and antimatter should exist in equal quantities, our experience in the world and our ability to avoid constantly being annihilated by everything we touch shows us that regular matter is winning by a very wide margin. How that came about is still a mystery, but it's one for which the clues are likely to lie in deeper and more detailed studies of the early universe, when that asymmetry first occurred.

Wherever we end up looking for data, in the quest for a theory of everything, we have two complementary approaches. One is to examine the phenomena we already see in nature that don't fit into established physical theories, so we can build new and better theories to explain them. The other is to just try to break the theories we have—write down hypothetical extreme cases that might not have been tested yet, and see if we can find a new way to look at the data that will show us if the theory still works there. A combination of these two approaches is pretty much always how we move forward in physics. It's how we went from Newtonian gravity, which works extremely well in everyday situations, to Einstein's general relativity. GR would be massive overkill for a block sliding down an inclined plane, but is absolutely essential for explaining the bending of light around extremely massive objects in space or the tiny shifts in the orbit of Mercury deep in the gravitational well of the Sun.

Newtonian gravity had to be replaced so we could move on to the superior general relativity; now it's general relativity's turn to be supplanted by the next big thing.

But GR has been so resistant to these efforts, we might end up having to rearrange the entire universe instead.

MAKING SPACE

There's a classic episode of *Star Trek: The Next Generation* in which, through a complicated series of events, Dr. Crusher ends up being the only person on the ship while it's trapped in some kind of weird hazy bubble. So many strange things are happening, including the sudden disappearance of the rest of the crew, and it's all at such odds with the readings of the sensors, her medical expertise tells her there's a very good chance she's hallucinating. But when her medical diagnostic fails to come up with any problem, she reaches the next logical conclusion: "If there's nothing wrong with me," she says, "maybe there's something wrong with the universe!" As it happens (and, apologies for the spoiler, but the episode aired in 1990; you've had three decades to watch it), she's absolutely right.

For a while now, some physicists have suspected that the incongruous weakness of gravity might be forcing them to a similar conclusion. Maybe there's nothing wrong with the strength of gravity. Maybe there's something wrong with the *universe* that's making gravity *seem* weaker than it really is.

What could make gravity seem weak? The solution might end up being surprisingly mundane. It's leaking. Into another dimension.

Here's how it works. As you're probably aware, we usually think of our universe as having three dimensions of space (east-west, north-south, up-down). In relativity, we count time as a dimension as well, and we talk about locations in a 4D space-time (a position in space and some moment on the past-future continuum). In the *large extra dimensions* scenario, there is another direction, or several, that we can't access. All of the space part of our spacetime is limited to a 3D "brane"—think membrane—and a larger space extends outside it in some new direction (or directions) our limited human brains can only conceptualize mathematically. I should also mention that the "large" in "large extra dimensions" is a somewhat misleading term. Generally, if our universe does have extra dimensions, it might be effectively infinite in our usual three dimensions but extend no more than a millimeter in the new directions. (Imagine a large sheet of very thin paper—it's technically a three-dimensional object even though two of its dimensions are much larger than the third.) But to a particle physicist, accustomed to measuring distances that make atoms look big, millimeters may as well be miles. Accordingly, we refer to the extra space outside our own brane as the "bulk."

In this scenario, particle physics and gravity still act fundamentally differently from each other, but not because of their inherent strength. The difference is that all the forces of nature in particle physics—electromagnetism and the strong and weak nuclear forces—are confined to live on the brane. To them, the larger, higher-dimensional bulk doesn't exist. But gravity is not so limited. Gravity acts directly on spacetime, and that includes the spacetime outside of our 3D brane. So the gravity produced by a massive object in our own space loses a tiny bit of its apparent strength by leaking into the bulk, like an ink mark fading as it seeps into a sheet of paper. The fact that the new dimensions are so small compared to our ordinary ones means that this leakage isn't really noticeable until you're measuring the gravitational effects of things at millimeter distances, which is extremely hard to do. After all, most of the time, if you're near an object and you step a millimeter farther away from it, you're not going to notice that your gravitational pull toward it has appreciably reduced.

Once you figure out how to make measurements on millimeter scales, though, you can test whether or not the decrease in gravity is what you expected from the standard equations. Going back to the ink-on-paper analogy, if you drop a gallon of ink on a sheet of paper, it'll still look like you have a gallon of ink. But if you measure it by the drop, you'll notice losing some when it soaks into the paper's fibers. If extra dimensions are the width of millimeters, and you can measure gravity changing even on those kinds of scales, then the amount of gravity you're losing to that extra dimensional bulk becomes comparable to the amount you're trying to detect. You'll see a drop-off of gravitational strength that's more rapid than what you'd predict from general relativity in a nonleaky space, and that will make it obvious that something isn't right.

So far, while we still don't have any alternative consensus on an explanation for gravity's weakness, we also haven't found any solid indication that this leakage is actually happening, despite getting better and better at measuring gravity at very small scales. As appealing as extra dimensions may sound from a theoretical point of view, their existence is still more in the category of intriguing possibility than a confirmed feature of our cosmos. And to a large degree, the original motivation for them has faded, as almost all the most compelling theories explaining gravity's weakness through leakage have been ruled out, since they predict alterations at levels we should have already detected. Still, we're continuing the search, because if extra dimensions do turn out to be real, they offer an entirely new view of gravity, and of the universe. Our entire universe being on a brane contained within a larger spacetime brings up the possibility that there are other universes out there, perhaps on nearby branes, which may be capable of influencing our own with their gravity. Even more dramatically, interactions between branes could provide a new scenario for the origin of our universe. And, ultimately, its destruction.

Enter: the ekpyrotic cosmos.

COSMIC APPLAUSE

The first time I encountered the ekpyrotic scenario for the origin (and fate) of the cosmos, it was at a very engaging physics lecture at Cambridge University by one of its originators, Neil Turok. The second time, it was in a science fiction story about aliens. It's not often that somewhat esoteric theoretical constructions developed to solve complex problems in early universe physics show up in fiction, so it was something of a novelty at the time. The story, "Mixed Signals" by Lori Ann White and Ken Wharton, tells of a series of strange events that ultimately appear to be linked to gravitational waves. Specifically: weirdly powerful gravitational waves too regular to be due to the usual suspects—black holes or neutron stars colliding. Eventually the protagonists figure out the waves are a signal from intelligent beings, sent through the higher-dimensional bulk from *another brane*. The authors even name-check the ekpyrotic model, explaining that in this theory our universe is just one of several 3-dimensional branes in a higher-dimensional space, in which gravity but nothing else can travel. And if gravity can traverse the bulk, gravitational waves could be an excellent inter-brane communication mechanism.

While the existence of other civilizations on nearby brane-universes was never technically ruled out as a possibility, the main purpose of the hypothesis was to explain the origin and destruction of *this* universe. Not long after the lecture and the sci-fi story, I ended up doing my PhD thesis on early universe physics with Paul Steinhardt, who worked with Neil Turok to come up with the ekpyrotic model. While I focused more on other theories of our cosmic origins, I encountered the ekpyrotic scenario on a regular basis at group meetings and discussions. (Somehow, aliens never really came up.)

Since those days, the ekpyrotic scenario has been revised and generalized, and the latest version doesn't include extra dimensions at all. But as often happens in science, a novel idea that may not work in the end can still spur on a different way of thinking about the problem—one that can lead us in a totally new (and hopefully better) direction. So let's start with the original idea. It does, after all, present us with the possibility of an intriguingly dramatic cosmic end.

The term "ekpyrotic" comes from the Greek for "conflagration," a reference to the fiery origin and ultimate death of the universe in this scenario. In the standard, non-ekpyrotic story, the beginning of the universe included a period of cosmic inflation,* which we discussed in Chapter 2. Inflation causes a dramatic stretching of the cosmos during the first tiny fraction of a second, after which the decay of whatever was responsible for the stretching (we call it the *inflaton field*)† dumps a massive amount of energy into the cosmos to set up the "hot" phase of the Hot Big Bang. In the original version of the ekpyrotic model, on the other hand, the early universe is heated up by a spectacular collision of two adjacent 3D branes, one of which contains what will later become our whole cosmos. After the collision, the two branes go their separate ways, slowly drifting apart across the bulk and expanding. But they will come

---

*That thing where a first-draft theory is dramatically redesigned but still useful applies to inflation too. The original version of inflation is widely considered to be a stroke of genius in spite of being, in the end, a total failure. It didn't work at all, and was totally revamped by other physicists within about a year. What its originators did exactly right was to propose a general class of solutions that became the spark of a firestorm of creative ways to finally make the Big Bang work. The revamped version, what we sometimes refer to as "new inflation," became the basis of the kind of inflation we all talk about today.

†Because we like to give particles and their associated fields names that end in "on."

back. The ekpyrotic scenario is a cyclic one, with creation and destruction of the cosmos occurring over and over again.

Personally, I find this whole thing makes much more sense if you employ the oldest tool in the physicist's toolbox: hand-waving.

Your left hand is our 3-brane: the 3-dimensional universe in which we live. (Obviously, none of this is to scale. This is hand-waving, after all.) Your right hand is another, "hidden" brane.* First, place your hands together, with your fingers closed, in a prayer posture. This is the moment of cosmic creation. This is the collision that sets off the primordial fire. Both branes are filled with dense hot plasma at this moment, an unimaginably intense inferno forging the first atoms and carrying the humming plasma waves that, on our brane, we'll later see as fluctuations in the cosmic microwave background light. Now, slowly separate your hands to a short distance apart, keeping them parallel, and spread your fingers. The branes have drifted apart across the higher-dimensional bulk, and the space in each brane is, independently, cooling and expanding in its own way. There's no inflation phase in this model, just a steady expansion after the collision. And they're not expanding into the bulk between them; they're extending in their own branes parallel to each other. On our brane, your left hand, this is the cosmos we see today. While we can't perceive our motion away from the other brane, we do see galaxies recede into the distance as the 3D space we live in expands, and our universe becomes more and more empty, heading toward a Heat Death. We don't know what's happening on your right hand, the hidden brane. Maybe there are civilizations there too, watching their own universe empty out as it traverses

---

* Each of these branes is designated in the official literature as an "end-of-the-world" brane, because it lies at the boundary of the space. This seems fitting.

an unseen void. Maybe it's a quiet, desolate place, where for whatever reason matter never learned to arrange itself into life. Maybe they have puppies who can talk. Unless we somehow detect a gravitational wave signal from the hidden brane, we may never know its true nature, or if it even exists.

Now, let your hands slowly approach again and then suddenly slam your hands back together. In this scenario, after the branes have drifted to their maximum distance and expanded, they are attracted back together, to bounce off each other again. That clap—the bounce—has destroyed everything on both branes, ended our universe, and created a new Big Bang. Both universes are back in the hot phase, filled with a plasma inferno, a chaotic state whose reborn space harbors little or no physical remnant of what it once may have held. Now separate your hands and do the whole cycle again. And again. And again. A braneworld* ekpyrotic universe is eternal, cataclysmic, cosmic applause.

ROUND AND ROUND AGAIN

Whether or not we really live on a braneworld, and whether or not there are other branes out there across some higher-dimensional bulk, are still open questions. The general idea of a cyclic universe holds some appeal, though, as it is one of a very few reasonable alternatives to inflation with any chance

---

* The term "braneworld" specifically refers to models in which there are higher dimensions and our observable universe lives on a 3D brane within a larger space. It's sort of a type of multiverse, but usually when people talk about a multiverse they're referring to something different, like regions of a larger (3D) space where the laws of physics might be different, or even the Many Worlds interpretation of quantum mechanics, which is another matter entirely. Any construction that allows there to be more to reality than our observable cosmic volume is a kind of multiverse theory.

of duplicating its successes.* What shapes the ekpyrotic model and inflation will eventually take are yet to be determined—the newest ekpyrotic models don't require branes at all, while some versions of inflation now do. The big difference between ekpyrotic models and inflation is that where inflation solves a number of cosmological problems by introducing a period of rapid expansion in the very early universe, the ekpyrotic model does it through *slow contraction* just before the bounce. In the case of the braneworld model, this is during the phase when the branes are coming together. Like inflation, an ekpyrotic model might be compatible with the distribution of matter we see in the universe today, and can potentially explain why our cosmos appears to be so very uniform and flat (in the sense of not curving back on itself or having some other complicated large-scale geometry). The fact that everything is weirdly uniform makes sense if the branes are huge and parallel before the bounce—it means that the bang can happen everywhere in the same way at the same time, with only some slight quantum fluctuations adding the necessary blips to set up the higher-density regions that will grow up to become galaxies and clusters of galaxies and all of cosmic structure.

As with inflation, however, a lot of theoretical details are still being worked out. The biggest is the question of what exactly happens during the bounce. Does a true singularity occur? Or does the bounce happen without the ultimate maximum density being be reached, allowing for the possibility that information of some kind could survive the event and pass on into the next cycle? The latest version of the model has very little contraction, so nothing like a singularity occurs. Instead of

---

*I am using "cyclic" and "bouncing" more or less interchangeably here, but a bouncing model doesn't have to be cyclic, in the sense that there could be only one "bounce"—a transition from some past long-lived pre–Big Bang phase to our current universe, which then dies out on its own without producing a new universe afterward.

using a collision between branes, the contraction in that model is driven by a *scalar field*, which is something similar to the Higgs field, or (possibly) to what we think might have caused inflation. That model does offer the tantalizing possibility that information might pass between cycles, and we could in principle someday see evidence of that.

Which brings us to the question of observational evidence. Since both the ekpyrotic model and inflation were designed to solve the same cosmological problems, it may take a little creativity to confirm or rule out either of them. Everything we've seen so far in the cosmos seems compatible with the standard inflation picture, but we haven't seen a smoking gun for it, nor have we seen anything that proves or kills off an ekpyrotic alternative. Arguments have gone back and forth for years about whether cyclic models are more or less *theoretically* appealing than inflation, but observationally it's still an open question. Some data to finally decide the issue would be extremely helpful here.

Our best bet might be finding evidence for *primordial gravitational waves*: large-scale ripples in space that would originate not from merging black holes or neutron stars but from the violence of the inflationary epoch, when the first seeds of cosmic structure were laid down by quantum wiggles in the inflaton field. If found, these would be as close as we're likely to get to the elusive smoking gun for inflation. Briefly, in 2014, the cosmology community lit up with excitement as the leaders of an experiment called BICEP2* announced that they'd seen evidence for exactly that. Looking at the polarization of the light from the cosmic microwave background, they saw what appeared to be twisting patterns that could only have come from gravitational waves distorting space during the epoch of

---

*The second iteration of the Background Imaging of Cosmic Extragalactic Polarization experiment.

primordial fire. These patterns were held up as a discovery so revolutionary Nobel Prizes were virtually guaranteed. After all, even setting aside the implications for inflation, they were a solid observation of gravitational waves (more than a year before LIGO saw its first black holes collide) *and*, because of the connection to quantum wiggles, they were the first evidence of any kind for the quantum nature of gravity.

Except, they weren't.

After just a few months, physicists and astronomers outside the BICEP2 collaboration independently analyzed the data and found that the pattern could be completely explained by something much more mundane: ordinary cosmic dust, in our own Milky Way galaxy. If primordial gravitational waves *had* been discovered, they would have been evidence against the ekpyrotic model, since that model doesn't include the inflationary universe-quake that could produce them. Unfortunately, their nondetection takes us back to square one. While inflation theory says that primordial gravitational waves must be produced, there's nothing in the theory that says they have to be *detectable*. The most popular inflation models give you substantial gravitational waves, but it's entirely possible to come up with one that produces a signal too weak to compete with the confusion of the cosmic dust.* So the fact that the dust got in our way doesn't prove that the inflation signal isn't there, any more than it proves that it is.

Still, we might get clues from other sources. We might find evidence for or against braneworlds in the search for extra dimensions, or we might finally get a hint of those primordial gravitational waves. Even ordinary gravitational waves could hold clues, either by showing us a signal that travels through

---

* Technically, depending on the model, you could also get some tiny tiny level of primordial gravitational waves in an ekpyrotic universe during the slow contraction phase. But they'd be WAY too small to ever show up in observations.

the bulk (via interdimensional aliens or not),* or by helping us to map out the structure of spacetime by, essentially, watching how it wiggles. According to some studies, data from black hole collisions have already put a damper on theories involving gravity leaking out into a higher-dimensional void. So far, all our measurements are consistent with a plain old boring universe with only three spatial dimensions.

Whether or not we find extra dimensions, the idea of a cyclic universe will likely continue to hold appeal as an alternative to inflation. One reason is the problem of entropy, the ever-increasing disorder in the universe that ultimately leads to a Heat Death. We can calculate the amount of entropy in our observable universe, and we can look back through cosmic history to determine what it must have been at early times if it has been steadily increasing over the lifetime of the cosmos. The result is that the universe must have started at a shockingly low-entropy—highly ordered—state when our own cosmic history began. This is a deeply uncomfortable idea for a lot of cosmologists. How did the entropy get set so low at the beginning? It's as if you walk into a room you're sure no one has ever been in before and you find rows and rows of dominoes lying on the floor, overlapping as if they've just toppled upon each other in sequence. How did they all get so carefully set up in the first place?

A major bonus in certain cyclic and bouncing models is that they offer an opportunity to attribute that low initial entropy to something that happened before the bounce. The latest update to the ekpyrotic model, developed jointly by Paul Steinhardt and Anna Ijjas, explains the early universe's low entropy by effectively taking all the entropy from a tiny patch of the pre-

---

* The idea that there might be matter on the hidden brane has been discussed in the literature, but as far as I'm aware, detecting black hole collisions across the bulk has not. Perhaps it would require too many levels of speculation for a serious study. But I think it sounds like fun.

bounce universe and setting that as the initial entropy of the entire observable universe today.

This new model (which is so new that it appeared on the scene during the writing of this book) has some significant advantages over previous versions of the ekpyrotic scenario. In particular, it doesn't require extra spatial dimensions or a singularity at the bounce. In fact, the contraction might be fairly mild—the reduction in size of the universe might be as low as a factor of two. The details are (obviously) complicated, but the basic idea is that what's really cycling is the mix of ingredients in the universe, and the way observers would perceive its evolution. As mentioned before, it's a scalar field filling the universe that drives the contraction/bounce, rather than a brane collision.

If this new cyclic model describes our universe, then in some far future epoch, we will start to see distant galaxies stop in their expansion and slowly turn around back toward us. It will look, at first, like the early stages of a Big Crunch, with the background radiation starting to heat up from "cool" to "this is not quite as cool" as the cosmos gets just a tiny bit more crowded. But just as we start to think maybe we should worry, we are, out of nowhere, suddenly and spectacularly obliterated when the scalar field violently converts its energy into radiation and starts the next Big Bang cycle of the universe.

Intriguingly, one aspect this brand-new hot-off-the-presses version of the ekpyrotic model shares with the old one is that rogue gravitational waves could be a kind of inter-universe signal. In the old version, it's conceivable some gravitational waves could pass through the bulk from another brane. In this one, since the cosmos never gets truly small during the bounce, gravitational waves might pass from one cycle to the next. These signals would be incredibly hard to find, but if they existed, they could present us with clues about a universe before our own.

Watch this space.

• • •

Of course, ekpyrotic models are not the only way to put some bounce in our cosmic step.

Roger Penrose, an early pioneer of modern cosmology who fundamentally changed the way we look at gravity in the universe, has his own proposal for a cycling cosmos, in which our Big Bang is born from a previous cycle's Heat Death. It involves piecing together the far future spacetime of one universe and the singularity at the beginning of another. Penrose has been, for decades, one of the most prominent voices in cosmology pointing out the seriousness of the entropy problem in standard early universe scenarios. And he does *not* think inflation does the trick. He told me recently, "When I first heard about it, I thought, well, that theory won't last a week."

Penrose's alternative model, called Conformal Cyclic Cosmology, conjectures that entropy works differently in the vicinity of singularities. If the conjecture is true, it implies that the entropy would be very low at the boundary between cycles from which our universe begins, and it doesn't require inflation. Penrose's model also contains the intriguing possibility that some imprint of the events that occurred in past cycles might appear in astronomical observations, showing up as features in the cosmic microwave background. In fact, Penrose and his collaborators have claimed that evidence for such features can already be seen in the data, though this has been met with skepticism. Whether or not these possible CMB hints will someday be seen as a compelling sign of a pre–Big Bang universe is yet to be determined.

Meanwhile, ekpyrotic-model co-developer Neil Turok has shifted focus to dive into a totally new model of the universe in which the Big Bang is merely a transition point. This proposal, developed by Latham Boyle, Turok, and their former student Kieran Finn, is motivated by taking symmetry arguments in particle physics to a cosmic level; it suggests that our universe and a time-reversed version of the cosmos meet at the Big Bang like two cones touching tip to tip. In a recent paper, they describe the picture as "a universe-antiuniverse pair, emerging from nothing." It's possible that cone-tip singularity might contain its own solution to the entropy problem, though the model and its details are (at time of this writing) still under development. Nonetheless, it makes some specific predictions for the nature of dark matter, and thus might be testable with upcoming experiments.

So where do we go from here? Was the Big Bang unique, or just a violent transition point? Will our cosmic existence be dramatically cut short by another universe coming down on us like a higher-dimensional flyswatter? Will data from cosmology or particle physics ever reveal the true nature of space-time? How close are we to finding out what our cosmic far future holds, and what new information do we need to answer the question, once and for all?

How will it all end?

Like everything in science, our understanding of the cosmos is a perpetual work in progress. But that progress has, over the last few decades, been extraordinary, and new insights are coming fast. Over just the next few years, humanity will gain new tools that will give us an unprecedented view of our cosmic history, allowing us to piece together the story of our origins and open new windows onto the Big Bang, dark matter, dark energy, and our trajectory into the future. In the final chapter of *this* story, we'll get a glimpse of what those new tools might show us, and how work on the cutting edge of physics is already pointing us toward a universe far stranger than we ever could have imagined.

CHAPTER 8:

Future of the Future

How big the hourglass?
How deep the sand?
I shouldn't hope to know, but here I stand.
Hozier, "No Plan"

In 1969, Martin Rees was not yet Astronomer Royal, Lord Rees, Baron of Ludlow. He was a postdoc cosmologist at Cambridge University, thinking about the end of everything, publishing a six-page paper titled "The Collapse of the Universe: An Eschatological Study," which he would later describe as "rather fun." In the introduction, Rees explained that while the observational evidence was still uncertain, it indicated that "the universe is indeed fated to collapse. All structural features of the cosmic scene would be destroyed during this devastating compression." Part of what made the paper fun to Rees was calculating that, in the coming collapse, all the stars will be destroyed by ambient radiation, from the outside in. Who wouldn't enjoy the thought of stars catching fire?

Despite Rees's arguments in favor of a Big Crunch, the data remained ambiguous for decades. Was the universe closed (recollapsing) or open (eternally expanding)? In 1979, Freeman Dyson, at the Institute for Advanced Study in Princeton, decided to explore the other side of the argument, saying, "I shall not discuss the closed universe in detail, since it gives me a feeling of claustrophobia to imagine our whole existence confined within the box." The open universe model was a pleasantly roomy alternative. In his paper "Time Without End: Physics and Biology in an Open Universe," he worked through quantitative predictions of what an open universe might mean for humanity, working out a method by which future beings might, through regulating their activity and entering periods of hibernation, avoid oblivion into the infinite future as the rest of the cosmos dissolves around them.* While most of the paper consists of calculations and theoretical discussions, the introduction contains some sharp words aimed at the physics mainstream for unfairly disdaining the whole endeavor of studying the cosmic end times. "The study of the remote future still seems to be as disreputable today as the study of the remote past was thirty years ago," he wrote, pointing out the scarcity of serious papers approaching the subject.† He continued with a cosmological call to arms: "If our analysis of the long-range future leads us to raise questions related to the ultimate meaning and purpose of life, then let us examine these questions boldly and without embarrassment."

I can't exactly say that cosmic eschatology has, after all this time, finally received its proper level of respect as an academic discipline. It's still rather rare to find papers in the physics literature that examine our ultimate fate with the same rigor and depth as they do our origins. But studies of both ends of the

---

* Unfortunately, the only kind of open universe model that will allow this is one without a cosmological constant, so even this tiny spark of hope seems to have been snuffed out by the current data.

† Amazingly, Dyson himself never submitted his own paper. It was submitted to *Reviews of Modern Physics* on his behalf by a friend, who didn't ask permission. Dyson told me recently, "I didn't think it was worth publishing," considering it not to be appropriate for the journal. "It's always a matter of opinion," he added.

timeline help us, in different ways, to examine the principles of our physical theories. Beyond the insight they might provide into our future or past, they can help us understand the fundamental nature of reality itself.

"By thinking about the end of the universe, just like with its beginning, you can sharpen your own thinking about what you think is happening now, and how to extrapolate. I feel like extrapolations in fundamental physics are essential," says Hiranya Peiris, a cosmologist at University College London. In 2003, she led one of the teams interpreting the first detailed view of the cosmic microwave background with the Wilkinson Microwave Anisotropy Probe (WMAP) satellite and she has since then maintained a position at the leading edge of observational cosmology. In recent years, she's set her sights on using observational data, simulations, and tabletop analogs to test some of the key elements of early- and late-universe physics like the creation of "bubble universes" in cosmic inflation and the mechanics behind vacuum decay. In studying all these questions, her motivation is the same. "I know this period needs to be understood. How what we're doing now will map directly onto those periods is still not clear, but I think we'll learn something about fundamental theory by doing this work."

We certainly have a lot to learn. Cosmology and particle physics are in an awkward position at the moment; both have, in some ways, been victims of their own success. In each field, we have a very precise and comprehensive description of the world that works extremely well in the sense that nothing has been found to contradict it. The downside is that we have no idea *why* it works.

The reigning paradigm in cosmology is called the Concordance Model, or ΛCDM. In this picture, the universe has four basic components: radiation, regular matter, dark matter (specifically "cold" dark matter, CDM), and dark energy in the form of a cosmological constant (denoted in equations by the

Greek letter lambda, Λ). The quantities of all these components are precisely measured, with the cosmological constant currently making up the largest slice of the cosmic pie. We have a good understanding of how these things have all varied over time as the universe has expanded, and we have an amazingly detailed description of the very early universe that includes a period of very rapid expansion called inflation. We also have a tried-and-tested theory of gravity, Einstein's general relativity, which in the Concordance Model is taken to be completely correct. In this picture, because the cosmological constant is currently dominating the evolution of the cosmos, we can straightforwardly apply our understanding of gravity and the components of the universe to determine our cosmic evolution. Doing this leads us unambiguously to a Heat Death in the far future. And that's that.

The problem with the Concordance Model is that the most important elements of it—dark matter, the cosmological constant, and inflation—are completely mysterious. We don't know what dark matter is; we don't know how inflation happened (or if it even really did happen); and we have no reasonable explanation for why the cosmological constant exists or why it takes a value that seems to fly in the face of what we expect from particle physics. At the same time, we haven't found anything in the data to contradict the model. No evidence that dark energy evolves in some way (which would go against a cosmological constant), no evidence that dark matter is anything experimentally detectable (and no evidence that it's not), and despite a century of putting it through the experimental ringer, no evidence for gravity behaving like anything other than Einstein's general relativity.

Andrew Pontzen, a colleague and coauthor of Peiris's (and my former officemate at Cambridge) works on theoretical aspects of dark matter, and has done some of the pioneering work to explain why dark matter takes the shape it does in galaxies. He contends that we have a very good understanding of cosmology, in the sense that our data line up extremely well with a picture that includes dark matter and dark energy, and that it seems unlikely anything will suddenly appear to change that picture. We know how much stuff is out there and how it behaves. On the other hand, we don't know how to connect either dark matter or dark energy, which together constitute 95 percent of the universe, to fundamental physics. "So in that sense we don't understand at all," he says.

Meanwhile, the view from particle physics is frustratingly similar. Back in the 1970s, physicists developed the Standard Model of particle physics to describe all the known particles of nature: the quarks that make up protons and neutrons, the leptons like the neutrinos and the electron and its cousins, and the so-called gauge bosons that act as go-betweens carrying the fundamental forces between particles (electromagnetism and the strong and weak nuclear forces). Notwithstanding some minor tweaks, like taking neutrinos from being strictly mass-less to being very, very light, the Standard Model has been fantastically successful, passing every experimental test thrown at it. It even predicted the existence of the Higgs boson—the final piece of the Standard Model puzzle. In the years since, nothing has been discovered in particle experiments that the Standard Model didn't tell us we would find.

You'd think this would be hailed as a triumph. The theory works! Everything is as we predicted!

Why aren't we sitting back and basking in our brilliance and success?

Because this is, in some ways, a worst-case scenario. As great as the Standard Model is at matching experimental results, we know that it, like the Concordance Model in cosmology, has to be missing some very important pieces. In addition to having nothing at all to say about dark matter or dark energy, it has some major "tuning problems"—places in the model where a parameter has to be set juuuuust right or else everything falls apart. Ideally, we should have some theoretical framework that tells us why a parameter is what it is. It's disconcerting when we find that the only reasons we have to set the parameter to that value are "otherwise bad things would happen to us" or, worse yet, "that's just what the measurement says."

For decades, there was hope on the horizon that we might be able to step seamlessly from confirming the important aspects of the Standard Model to finding the edges of its validity and making new discoveries with whatever model we found to replace it. In the 1970s, a model known as supersymmetry (SUSY, for short) was proposed to fix some of the theoretical niggles of the Standard Model by hypothesizing new mathematical connections between different kinds of particles and explaining the confusing structure of the Standard Model and its parameters. It came with a tantalizing promise too: a whole slew of new particles ("supersymmetric partners" of the Standard Model set) that might be produced in particle collisions just a *little* more powerful than what could be achieved by colliders at that time. SUSY has also been widely held up as a stepping-stone toward string theory, the leading idea in the quest to bring gravity and quantum mechanics together into a unified whole.

Unfortunately, despite working for decades to improve and upgrade the LHC, we've seen no sign of supersymmetry's promised particles. Some physicists still hold out hope for SUSY by proposing tweaks that would make the new particles harder to find, but at some point the tweaks become so extreme that SUSY has just as many theoretical problems as the Standard Model. And the signal just isn't there. Now and then, some quirk of the data will produce a whirlwind of excitement as physicists rush to explain why there are a few more events in a particular detector channel than expected. But so far, none of these blips has turned out to be more than a statistical fluke destined to fade away in the next data release.

I spoke to Freya Blekman, an experimental physicist who searches for beyond–Standard Model signatures in LHC data, about the current conundrum. "I've been in the field twenty years now and I've seen my share of excesses come and go, and I've also seen my share of popular models come and go," she said. "Depending on who you speak to, there are people who are disillusioned . . . people have been telling them for a very long time that they should be seeing something. And what the experiments see is only the Standard Model." From her perspective, though, the disillusionment is misplaced. Not because people are missing hints that are really there after all, but just because there was never any guarantee that anything new would be found with these experiments.

Still, the lack of direction from experiments can be troubling—enough to push some researchers out of particle physics entirely, and into cosmology. One of those is Pedro Ferreira, a cosmologist at Oxford University who switched from quantum gravity to cosmology during his PhD and who now studies the cosmic microwave background and general relativity in astrophysics in the hope that they might provide some better insights. "There hasn't been anything revolutionary that particle theory has done which has led to observational results since 1973," he says. There have been lots of new theoretical ideas, and some of them are very appealing, but without clear experimental evidence for something beyond the Standard Model, it's hard to know where to go next, or which of the various proposals are likely to be right. "There's all this beautiful stuff that's come out. But have we solved the problem of quantum gravity? I don't think so. And the problem is, how would we know if we'd solved it?"

Fortunately, no one is giving up hope. I spoke with dozens of cosmologists and particle physicists about where this whole thing is going (where by "whole thing," I mean both theoretical physics/cosmology and the actual universe), and while there was no agreement about the optimal approach, there were a few common themes. One was diversification: whatever big multinational experiments or observational programs we decide to invest in, it's important to diversify our approaches and come up with ideas that will give us new perspectives on these old problems (which goes for the theory side as well as the data-taking side). The other was the importance of continuing to get as much new data as we can, and to analyze it in every way possible.

Clifford V. Johnson, a theoretical physicist at the University of Southern California, works on string theory, black holes, extra dimensions of space, and the subtleties of entropy. He is about as deep into pure theory as anyone I know, and he is *very* excited about data right now. "My feeling is that we are maybe lacking a good sort of single idea, but we are not lacking in huge sources of data," he said. "And that reminds me of the immediate pre-quantum days, right?" In those days, theory was booming, with lots of half-formed ideas about the structure of atoms and nuclei, though none were all that compelling. "But we just then got all this wonderful data that began to eventually take shape. I don't see why that can't happen again. Looking at the history of science, that's how this works."

So let's talk about data. What we're looking at, and how, in both cosmology and particle physics. What it might tell us about both the physics of the universe today and how it's all going to come to an end in the future. And then we'll check back in with the theorists. Because some of the ideas they're talking about right now are absolutely wild.

TOUCHING THE VOID

If we want to learn anything about the far future of the cosmos, we'd better address the giant invisible ever-expanding killer elephant in the room: dark energy. When the accelerated expansion of the universe was discovered in 1998, the new paradigm placed us squarely in the path of a dark-energy-dominated future: one in which the cosmos gets progressively emptier, colder, and darker until all structure decays and we reach the ultimate Heat Death. But this is just an extrapolation, one that's predicated on dark energy being an unchanging cosmological constant. As we've seen, if whatever is responsible for cosmic acceleration falls into the category of phantom dark energy, or if it somehow changes over time, the implications for the cosmos are drastically different.

Unfortunately, as far as observations are concerned, dark energy doesn't give us a lot to hold on to. It is, as far as we can tell, invisible, undetectable in laboratory experiments, completely uniformly distributed through space, and only really noticeable at all by its indirect effects over scales much larger than our galaxy.

Generally speaking, there are two things we can measure. The first is the expansion history of the universe, which at the moment we study primarily by looking at very distant supernovae and figuring out how fast they're receding. The other is the history of the formation of structure, where by "structure" we are generally referring to galaxies and clusters of galaxies, because all the little things like stars and planets are just annoying details if you're a cosmologist. Measuring this is a bit less straightforward, but also allows for a lot of creative uses of massive piles of data. The trick is to get images and spectra of as many galaxies as possible, over a giant volume of space (and a large patch of cosmic history), and use statistical methods to infer how all that matter came together over time. Together, these two kinds of measurements can tell us how the space-stretching properties of dark energy have affected the universe as a whole and how much it's impeded the efforts of matter to clump together and form things like galaxies and clusters and us.

When you only have two things you can measure to determine the WHOLE FATE OF THE UNIVERSE, it makes sense to invest a lot in measuring them very, very well. There's been a surge of interest in the last couple decades in new telescopes and surveys with "dark energy" prominently featured in their science cases. Some are designed around the promise of how well they can use expansion and structure growth measurements to determine the dark energy equation of state parameter w (discussed in Chapter 5). If w = -1 exactly, now and in the past, we have a cosmological constant, and if it's measurably different by any amount at all, we have a lot of Nobel Prizes. But even if you don't care about dark energy, or if you subscribe to the pessimistic view that we're fated to forever just narrow in on a garden-variety cosmological constant, dark energy surveys tend to be popular among astronomers of all stripes by doubling as all-purpose galaxy-gathering missions.

The upcoming LSST (Large Synoptic Survey Telescope), recently renamed the Vera C. Rubin Observatory (VRO), is a fantastic example. An 8.4-meter telescope on a high-desert mountain in Chile, VRO will take images of a few million supernovae and 10 BILLION galaxies, piecing together new images of the whole southern sky every few days. That kind of repeated coverage is great for supernova studies because it'll let us see the rise and fall of the brightness of each supernova over the several days during which the explosion is visible. But it's also great for studying galaxies, because it means you can stack up images night by night and see fainter and more distant galaxies than any other survey of its kind.

(As an aside, I recently attended a conference session about Planetary Defense in which the speakers were discussing the kinds of observations you need to spot potentially hazardous asteroids that might be on a collision course with our fragile little planet. The VRO will, at least for the southern sky, revolutionize our ability to pick these things up early, which might make it easier to find ways to stop them. I get a kick out of the idea that by attempting to understand the dark energy that will eventually destroy the universe, we might have a better chance of, on a much shorter timescale, saving the world.)

Whatever else its uses, the cosmological value of the VRO can't be overstated, if only because having massive piles of exquisite data gives us a very good chance of finding something new and surprising. According to Peiris, the VRO will be a game changer. "We are looking at the universe in a different way than what's been done before," she says. "Any time we've looked at the universe in a way that we haven't before, we learn new stuff."

VRO isn't the only new observational program to get excited about. There are a slew of other new telescopes and surveys coming up, each of which is poised to show us the cosmos in ways we've never seen before. Some of the most hotly anticipated are a class of new space telescopes like the James Webb Space Telescope (JWST), Euclid, and the Wide Field Infrared Survey Telescope (WFIRST), which will take deep images and spectra with infrared light, helping us to see galaxies so far away that their light has been stretched out of the visible part of the spectrum altogether.

Even cosmic microwave background observatories are getting in on the dark energy game. We saw in Chapter 2 how studying the CMB can tell us about the early universe and the origins of cosmic structure. At the time the CMB light was emitted, dark energy was completely unimportant in the universe, its effects totally swamped by the extreme densities of matter and radiation. So it may be surprising that CMB observations give us any insights into how dark energy is acting today. The trick is that all the cosmic structure we want to study—every galaxy and cluster of galaxies—is *between* us and the CMB, and each one of those objects distorts the space it's in just a little bit with its gravity.

Imagine you have a snapshot looking down into a clear-water pond at the pebbles below. Even if you don't know exactly where every pebble should have been placed, or all their exact shapes, you could probably tell the difference between very still water and water that has some ripples in it by noticing distortions in how the pebbles look, because you have a sense of what pebbles should look like in general. In a similar way, we understand the cosmic microwave background so well that we can see, at least in a statistical sense, the tiny distortions in its light due to all the stuff between here and there. This is called CMB lensing, and it's a fantastic tool for studying the growth of cosmic structure. New CMB observatories will help us refine the method, but we've already used CMB lensing to make a map of ALL OF THE DARK MATTER IN THE OBSERVABLE UNIVERSE. Granted, the map is an extremely low-resolution, blurry kind of map, like a map of the world reproduced from memory with fingerpaints, but still, it's pretty impressive that this is a thing we can do at all.

Renée Hložek, a cosmologist at the University of Toronto, uses the CMB and galaxy surveys to better understand our cosmological model, with a particular interest in dark energy and the universe's ultimate fate. She points out that combining data between things like VRO and new CMB observatories will become especially powerful as each data set improves. Using a technique called cross-correlation, we can take what we know about the positions of individual objects from galaxy catalogs and compare that with what we know about the largest-scale distribution of matter from CMB lensing. This can give us more precise results that make it harder to miss any deviations from the Concordance Model. Alternative theories that use changes in gravity to mimic the effects of dark energy will look very different in the combined data, Hložek says. "Basically, I think we'll run out of places to hide."

What other cool things can you see if you have images of

billions of galaxies? A big one is strong gravitational lensing, in which a galaxy or cluster of galaxies is distorting the space it's in so much that the light from an object directly behind it gets split into multiple images, or spread out as an arc of light encircling it. Think of looking at a candle through the base of an empty wineglass—the curved glass spreads the light out in broad arcs or a circle instead of showing it to you as a single flame. When a gravitational lens does this, the individual images follow different paths through the distorted space. That means that if, for example, a supernova goes off in the lensed galaxy, it might be seen in one of the images *before* it shows up in another, because the light making up the second image took a longer path to get to us.

Aside from being a fabulous party trick,* time delay measurements like this give us a new way of measuring the expansion rate of the universe, since the distances involved are so large that the expansion becomes an important factor in the calculation. And we desperately need new ways to measure the expansion rate, because our current methods are giving us weirdly different answers.

As you'll recall from Chapter 5, measuring the expansion rate (also known as the Hubble Constant) using supernovae gives us one number, and measuring it via the CMB gives us another. A slew of other measurements have failed to resolve this contradiction, generally falling on one side or another. (A very recent result found something in between, but in a way that, unhelpfully, didn't agree with either side.) Gravitational lensing time delay measurements might be a way to resolve the problem, because with the VRO, the number of these systems we can use will go from a few to hundreds. Gravitational

---

* "See that star there? That star is going to EXPLODE in one year. Plus or minus four months. Just watch, you'll see." (Adapted liberally from Treu et al. 2016, *The Astrophysical Journal*.)

wave measurements from instruments like LIGO (discussed in Chapter 7) could give us insight here too, and in the next decade or so might reach the precision necessary to finally settle the question.

THE VIEW FROM LEFT FIELD

One of the things I love about cosmology is how much it requires thinking creatively, trying to approach the physics of the universe from a totally new direction. This doesn't mean fully unconstrained flights of fancy. You can't just randomly make stuff up. But what you can (and must) do is constantly find new ways to look at problems to wring a little more insight out of whatever data the universe has to offer.

This kind of creative thinking becomes especially important when we're faced with a conundrum like "How do we improve on Concordance Cosmology or the Standard Model?" Everything we've tried so far has been frustratingly consistent with predictions; where are we supposed to find clues leading us to new models if we can't get something in the current model to break?

Clifford Johnson is optimistic, and points out that this lack of clear direction might be good for us. "I don't have a thing I can point to and go, 'This is the future!'" he told me. "I just feel the diversity of things that we've been driven to do . . . is probably somewhat healthy."

So, we're branching out. There are radio surveys attempting to illuminate the cosmic Dark Ages between the time of the CMB and the epoch of the first stars, in the hope that some departure from Concordance Cosmology might reveal itself more starkly. There are new kinds of gravitational wave detectors that rely on techniques as different as quantum interference between atoms and combining signals from pulsars.

These might, in an indirect way, bring us information about the behavior of black holes or the physics of the early universe. Experiments looking at new ways to find dark matter might show us how to expand the Standard Model of particle physics, or shift our thinking in cosmology. Studies of the polarization of the CMB could show us signatures of cosmic inflation that completely change our understanding of the early universe. Or, the lack of such signals could motivate more studies into inflation alternatives like bouncing cosmologies. Laboratory experiments studying alternative ideas about the energy of the vacuum might finally solve the problem of dark energy, if it's not a cosmological constant after all. It may even be possible, with observations spanning decades, to measure the expansion of the universe *directly* by staring at a distant source for so long that its apparent speed away from us changes.

Pedro Ferreira is also optimistic about this diversity of approaches. "I think it all might look quite specialized and bitty," he says, but having a huge number of people suddenly individually racking their brains to come up with something new could be exactly what we need. "Out of that explosion someone might have an idea. 'Oh! This is the way to figure out the future.'"

How long such a program will take is another question. If we're just trying to distinguish between a cosmological constant and some other form of dark energy, we literally have all the time in the world, and then some. There's really no theory out there in which dark energy can destroy our planet before our own Sun does the job.

But vacuum decay is another matter. The Standard Model of particle physics, the very same one that has passed every experimental test we've come up with, places us in a precarious position on the edge of total universal instability. How likely this is to be an actual risk, or a quirk of the extrapolation of an incomplete theory, depends on who you ask. (For the record,

I asked several experts and I got answers ranging from "it tells us our theory is wrong" to "the risk is really tiny" to "maybe we've just been lucky so far." Take that as you will.) In any case, if we want to be able to say something more reassuring than "It's useless to worry because you are not going to feel any pain,"* we're going to need a very specific kind of data.

Fortunately, we have a pretty good idea where we can get it.

DISCOVERY MACHINES

There is no place on Earth with a more persistent, if wholly undeserved, association with the destruction of the cosmos than CERN. Best known as the home of the Large Hadron Collider, CERN is a sprawling campus of laboratories and office buildings covering about six square kilometers that straddles the France-Switzerland border near Geneva. It's essentially an oddly specialized little border town, complete with its own fire department and post office, alongside laboratories and machine shops and a bona fide antimatter factory. Physicists at CERN have been accelerating and smashing protons since the 1950s, long before the LHC was built, carrying out increasingly complex and sensitive experiments to examine the nature of subatomic particles by obliterating them against each other. These kinds of experiments helped us to create the Standard Model of particle physics, and more than fifty years of continued experiments have failed to find any cracks in it wide enough to stick a new particle through.

But CERN keeps trying. And not just because smashing things is, admittedly, quite a lot of fun.

The name of the game in particle colliders is energy. Throw-

---

* Thanks, Madrid-based theorist and CERN Scientific Associate José Ramón Espinosa. Very helpful.

ing particles at each other faster means the eventual collision is at a higher energy, and the higher energy your collisions, the larger the swath of possible new physics you can reach. You can think of collision energy as legal tender, to be exchanged, via $E = mc^2$, for particle mass. If the total energy in a collision is higher than the equivalent mass of the particle you're trying to create, then as long as your theory allows *any* kind of interaction between that particle and the ones you've smashed together, you have a chance of creating that particle. Extensions to the Standard Model tend to involve particles that are significantly heavier than those we've detected so far, which means that we need to reach higher and higher energies to find them. But even when you've reached the right energy threshold, it takes more than one creation of a particle to get a meaningful, statistically significant signal. The Large Hadron Collider had to run for years, smashing countless trillions of protons,* before it had collected enough data to say with acceptable certainty that a Higgs boson had been found.

It's this constant push into the energy frontier that leads to CERN's unfortunate reputation as an existential menace. The thinking goes that if humanity has never before seen so much energy concentrated into one place, *who knows what could happen?* Some of the concerns include the unsettling scenarios we've discussed in previous chapters, like the creation of little black holes, or the triggering of catastrophic vacuum decay. Fortunately, in every disaster scenario presented so far, we can easily set aside the worries based on the fact that the LHC is hardly even a blip compared to the particle-obliterating violence going on all around us in the universe. But in the minds of certain especially fretful nonphysicists, not every worry is as well defined, or as easily assuaged, even though the LHC

---

* Probably closer to 10^{15}, but I have a moral objection to the word "quadrillion."

has operated totally harmlessly for over a decade. By the time I visited CERN in February 2019, internet jokes about the LHC tearing open a portal into another dimension, or shifting the universe into "the bad timeline," seemed about as prevalent as ever.

The CERN campus itself is not, for the most part, an especially impressive place. Once you get past the glitzy public reception lobby, it has the feel of a slightly run-down industrial installation, with a mishmash of low, drab, 1960s-era buildings with dark metal-shuttered windows. Each prominently numbered building houses its own lab or research group, the offices labeled with temporary paper nameplates to accommodate the constantly shuffling scientific staff. Across the entire campus, the physicists permanently employed by CERN number less than a hundred, with the rest of the labs and offices occupied by the thousands of visiting researchers from around the world, spending anywhere from a week to a few years carrying out the intense on-site work necessary to keep large-scale experiments running. Walking down the long dim hallways of one of these buildings, you might forget you're at the most famous experimental facility in the world, and imagine yourself in the physics department of any ordinary university, peeking in at grad students and postdoctoral researchers tapping away at laptops, or scribbling equations and work schedules on whiteboards.

When you see the experiments, though, that illusion of normalcy is swiftly and permanently broken.

My own visit to CERN was divided between the organization's two extremes. On some days, I was quietly ensconced in a bright second-floor office in the theory department, reading papers and taking breaks in the tea room to sketch out equations and chat with the other theorists about vacuum decay and my own research on dark matter. On other days, I was wearing a hardhat, 100 meters underground, standing on a metal walkway and gawking at a 25-meter-tall heavily instrumented cylinder of unimaginably complex machinery. The experiments at CERN are some of the most advanced and precise machines ever created by humanity, designed and built by teams of thousands over decades in order to tease out tiny changes in the motions and energies of particles that decay within microseconds. Meanwhile, theorists try to extract from equations of comparable but abstract complexity the implications of these experiments for the nature of space and the cosmos itself. It's a heady place.

It is also, however, an intensely bureaucratic place, being an institute governed by international treaties and run by a coalition of twenty-three different countries while hosting researchers from every corner of the planet. This kind of cooperation is necessary for an effort of such magnitude and expense, but the upshot of CERN's organizational structure is that the future of the facility and of any new experiments depends as much on international politics as on any scientific considerations. During my visit, the hot topic at the cafeteria was not some exciting new experimental result, but a series of back-and-forth newspaper editorials debating the merits of CERN's proposal to build the so-called Future Circular Collider (FCC), a particle collider so big that the 27-kilometer LHC would become merely a pre-accelerator to bring the protons up to a speed where they could begin to circulate in the FCC ring. The FCC could reach energies of 100 TeV, which is about an order of magnitude higher than what's currently possible at the LHC.

As Freya Blekman pointed out to me during my visit, these experiments take decades to set up, and the data from current experiments can take equally long to analyze, so discussions of the next experimental direction have to happen now. The kind of data we are already getting with the LHC and its upcoming upgrades will take us ten or even fifteen more years to fully analyze. "So this is the time to decide," Blekman says. "What do we want? Do we want an electron-positron collider? Should it be linear? Should it be circular? What are the pros and cons of each? Do we want to directly go to a higher-energy proton-proton machine?"

The arguments for and against future colliders, especially the ambitious FCC, can get rather heated. Even if you set aside the cost (around 10 billion euros at a minimum), debates remain around the promise—or lack thereof—that a bigger collider will find new particles. It may be that the elusive "new physics" we're searching for only shows up at energies so high that even gargantuan machines like the FCC have no hope of ever reaching them. Or it may be that just focusing on increasing the energy puts us on the wrong track entirely, and there's some clue about new physics hiding in another regime that we have yet to explore, perhaps even in data we already have.

Researchers I spoke to at CERN were adamant that increasing energy is essential to move us forward, even if only to better understand the Standard Model. Which does, after all, present us with the specter of vacuum decay. If that Sword of Damocles is going to be hanging over our heads, it would be nice to know exactly what it's doing up there.

André David, an LHC researcher in the Compact Muon Solenoid (CMS) collaboration who hosted my visit to the detector, pointed out that answering this question is a key motivation for the FCC and experiments like it. "One of the reasons why people are saying, 'Oh we should go for the hundred-TeV collider' is that then you actually get a chance at nailing this thing down."

As David points out, we already have a puzzle on the table: the nature of the Higgs field, and its (and our) fate. The data we've already obtained, and are working to analyze, could begin to trace out the nature of the Higgs in more detail, but with a new collider, we might finally answer the question of what the instability that threatens us with vacuum decay really means.

As we discussed in Chapter 6, the Higgs potential is the mathematical structure that determines how the Higgs field evolves, and, importantly for us, whether it will send us all to our doom. It is, in a real sense, the holy grail of particle physics. But with current theories we have very little handle on what it looks like. Based on our current understanding, its shape depends sensitively on the competing influences of several different hard-to-calculate aspects of the Standard Model, and if some higher-energy theory exists, this could completely change the picture.

Some researchers I spoke to, including CERN theorist (and leading supersymmetry advocate), John Ellis, suspect that the apparent instability of the Higgs is not really an existential threat, but rather a sign that there's something about the theory that we don't understand.

José Ramón Espinosa, a theorist studying vacuum decay, hopes to find ways to better understand the Higgs potential and what our precarious placement on the knife edge of stability might mean, without simply waiting for a true vacuum bubble to show up.* "There's no reason for the potential to be like this," he says. "We live in this very, very special place. So for me this is kind of intriguing; maybe this is trying to tell us something." The key to understanding the Higgs potential ultimately depends on what are called *running couplings*—the interactions between particles and fields and how they change with higher-energy collisions. "This might be one of the main messages of the LHC if we don't find anything else," Espinosa says. "Of course, if the LHC finds new physics, then most

---

* This method is especially undesirable, Espinosa points out, since it "will not teach us anything, because we won't even see it coming."

likely this is going to interfere with the running of the couplings. Then anything can happen. Maybe the potential is stable, maybe it's even more unstable. We don't know."

In addition to the small (but important!) point of determining the fate of the cosmos, a better understanding of the Higgs field could show us how mass works, or why the fundamental forces manifest with the strengths we measure. It could even point the way toward a theory uniting the forces, or help us to understand quantum gravity.

Having some kind of guidance from observations or experiments on how to improve on Concordance Cosmology or the Standard Model would be very helpful. Because over on the pure theory side of things, things are getting very, very weird.

THROUGH A GLASS DARKLY

I recently came across an old black-and-white photo of Paul Dirac, Nobel Laureate and pioneer of quantum mechanics, standing on the grounds of Princeton's Institute for Advanced Study with an axe slung over his shoulder. During his many visits there from the 1930s to the 1970s, he was known to wander through the woods behind the institute, clearing new paths for the resident theorists to walk and talk and think about the nature of reality. My own guide through those same muddy trails was Nima Arkani-Hamed, which seems appropriate, because he is a theorist determined to take an axe to our current understanding of quantum mechanics, and to the entire notion of spacetime itself.

Arkani-Hamed has been working on a way of calculating the interactions between particles using a completely new framework, one that starts from a kind of abstract mathematics in which space and time are not, strictly, included. The work is still in its early stages, and so far applies more to certain idealized systems than to experimental results. But if it works out, the implications could not be more mind-blowing. "What we're seeing is just, it's in baby, baby toy, toy, toy examples, right? You can justifiably use as many diminutives as you want on what's actually been accomplished, and I would be entirely sympathetic," he tells me. "But for what it's worth, there is starting to be one or two examples of actual concrete physical systems not so far from what we see in the real world where we can actually figure out how to describe them without either spacetime or quantum mechanics." I tell him I'm trying to wrap my head around what it means to live in a universe where space and time aren't real. He laughs. "Join the club."

Before you dismiss the idea as eccentric-theorist hyperbole, I should point out that Arkani-Hamed is not the only one talking like this. "I'm sure you've heard this from many people," Clifford Johnson tells me, nonchalantly, a few months later, "but I think we're getting better at realizing one of the things we've been saying in string theory for a long time, which is that spacetime isn't fundamental."

Oh, yeah. That small detail. Sure.

Johnson's approach to the question is a bit different. There are some intriguing hints in quantum gravity theories of unexpected connections between physics on small and large scales, in ways that don't make sense in our usual thinking about how spacetime works. A simplified explanation might be that if you imagine doing experiments in a hypothetical kind of space that has a certain radius, let's call it R, the results of that experiment would look exactly like the results of the same experiments in a much smaller space, with a radius equal to 1 divided by R. In string theory, this is called T-duality, and it's a weird enough coincidence that it seems like it *has* to be telling us something deep. "If you ask people about this question," Johnson says, "the answer that people would give is that in some sense, none of it is real. In the sense that by undermining large and small, what you're really doing is undermining the whole business of spacetime in the first place."

Some theorists have tried to reassure me. Sean Carroll, a cosmologist at Caltech who's interested these days in the underpinnings of quantum mechanics, thinks we are all being a little rash to dismiss spacetime as not strictly real. "It's *real* but not *fundamental*," he tells me. "Just like this table is real but not fundamental. It's a higher level of emergent description. That doesn't mean it's not real." Basically we shouldn't get too hung up on this because it's not like spacetime isn't there, it's just that if we really understood what it was made of, it would look, at a deeper level, like something else entirely.

This does not, in fact, reassure me.* As a physicist I always try to maintain some level of dispassionate reserve when it comes to my subject, but the notion that spacetime is only real in the sense that it is something we can talk about and sit on but not in the sense of being what the universe is *actually made of* still makes me feel like it might collapse underneath me at any moment.

Whether or not this has relevance to how or when the universe will end is still an open question. However real spacetime is or isn't, we all live there, and what happens to spacetime is bound to affect us. But if thinking about emergent spacetime or new formulations of quantum mechanics leads us to some deeper fundamental theory, it might drastically change our outlook. Maybe, as Johnson suggests, connections between large and small scales could imply a new fate for the cosmos. Or maybe, if we were able to revise quantum mechanics, we would finally find an explanation for dark energy. Even if we settle on a cosmological constant and a Heat Death future,

---

* Another thing Sean Carroll pointed out to me is that if his interpretation of quantum mechanics is correct, there are countless copies of ourselves in parallel universes that are at this very moment succumbing to vacuum decay. So he was probably never really going to be the best place to turn for relief from existential crisis.

according to Arkani-Hamed, we will still need a major shift on the theory side to be able to talk about what quantum fluctuations might do then, in terms of Boltzmann Brains or Poincaré recurrences. "In my mind it's very unlikely that all these things are explained and understood within the framework of quantum mechanics," he says. "I think we need some extension of quantum mechanics to help us talk about it."

To what extent there even exists an explanation for the nature of our universe is an open question too. In the last decade or so, physicists have been grappling with the concept of the *landscape*—a theoretical multiverse of different possible spaces which could have drastically different conditions from our own. If such a landscape really exists, it could mean that the properties of the space we live in are merely environmental, rather than being set by some deep principle we haven't been clever enough yet to find. This kind of multiverse can arise out of certain versions of inflation where new bubble universes inflate out of some eternal pre-existing space forever. "The idea that we're the unique solution of the world does not seem right to me," says Arkani-Hamed. "But on the other hand, when you try and make sense of the landscape and eternal inflation and all that stuff, it's such a morass that I think that the whole conception of the problem is wrong to begin with." Even with a landscape of possible universes, the basic problem still remains. "These questions about how to apply quantum mechanics to cosmology have been there almost from day zero. They're not new. They were very difficult fifty years ago; they're very difficult now."

"I very firmly believe what we should be doing is actually just retracing our steps," says Neil Turok, a cosmology theorist who has been looking at alternatives to cosmic inflation and who spent many years as the director of the Perimeter Institute for Theoretical Physics in Canada. "Go back, rewind fifty years and say, 'Guys, we're building on sand.'"

THE LONG VIEW

There's a famous equation in astrobiology called the Drake Equation. In theory, it's a way to calculate the number of civilizations in our galaxy with whom we might be able to communicate. All you have to do is input the number of stars, the fraction of those with planets, the fraction of those with life, the fraction with *intelligent* life, and so on, and in the end you get the number of messages you should expect on your interstellar voicemail. Of course, many of these input numbers are, at least with current data, completely impossible to determine, which means that the final answer isn't meaningful. The thing that's useful about the Drake Equation is that it makes us think about our *assumptions* about extraterrestrial life, and to figure out what we do and don't know about this whole question.

Talking with Hiranya Peiris, it occurs to me that contemplating our ultimate cosmic destruction might be much the same. I suggest to her that perhaps we're doing a calculation where the final number doesn't matter, but the calculation does. "The number doesn't matter," she agrees, "but the exercise of thinking through the different options on the table, I think, is good." And the implications of this thought experiment might ultimately pay off. "It could lead to some cool way to test between the hypotheses that doesn't wait for seven billion years."

How long *do* we have to wait for a breakthrough? We don't (and can't) know. We're exploring off the edge of the map now. Clifford Johnson is very optimistic that we're heading toward a better, deeper understanding of physics, but he acknowledges the caveat. "It might be that we go for a couple hundred years gathering all of this data before we see the signal and then we go back and realize that, oh, it was there staring us in the face all along. That's an annoying possibility. But for questions as big as the ones we're trying to answer, I feel that that's okay. Why need it be of the length scale of a human lifetime?"

In the meantime, we'll continue on, making new paths through the woods to see what we might find hiding there. Someday, deep in the unknown wilderness of the distant future, the Sun will expand, the Earth will die, and the cosmos itself will come to an end. In the meantime, we have the entire universe to explore, pushing our creativity to its limits to find new ways of knowing our cosmic home. We can learn and create extraordinary things, and we can share them with each other. And as long as we are thinking creatures, we will never stop asking: "What comes next?"

CHAPTER 9:

Epilogue

"But if nothing we do here has any guarantee of lasting, if even the best gestures have only a slim chance of outliving us, is there any reason not to just give up?"

"Every reason in the world," Rudd said. "We're here, and we're alive. It's a beautiful evening, on the last perfect day of summer."

Alastair Reynolds, *Pushing Ice*

Martin Rees isn't building any cathedrals.

We are sitting in his office at Cambridge University's Institute of Astronomy on a sunny June morning, and he is telling me that humanity as we know it will be forgotten. "In the Middle Ages, the cathedral builders were happy to build a cathedral that would be there for more than their lifetime, because they thought that their grandchildren would appreciate it and would live lives like them. Whereas I don't think we have that." Rees is no stranger to far-future speculation, having written books about the future of humanity and all the different ways we might accidentally doom ourselves. According to him, evolution, in the cultural and technological sense, is accelerating so fast that whatever the dominant intelligence is in the next few hundred or thousand years, we can't predict what it will be like. But we can be sure that it won't care about us. "I think to leave a legacy for a hundred years is a bolder ambition now than it would have been for our ancestors," he says.

"Does that upset you?" I ask him.

"It upsets me very much. But why should the world be made the way we like it?"

It's impossible to seriously contemplate the end of the universe without ultimately coming to terms with what it means for humanity. Even if you take the position that Rees's view is overly pessimistic, there has to come a point in any time-line with a finite extent where our legacy as a species just . . . stops. Whatever legacy-based rationalization we use to make peace with our own personal deaths (perhaps we leave behind children, or great works, or somehow make the world a better place), none of that can survive the ultimate destruction of all things. At some point, in a cosmic sense, it will not have mattered that we ever lived. The universe will, more likely than not, fade into a cold, dark, empty cosmos, and all that we've done will be utterly forgotten. Where does that leave us now?

Hiranya Peiris sums it up in one word: "sad."

"It's very depressing," she says. "I don't know what else to say about it. I give talks where I mention that this is probably the fate of the universe, and people have cried."

It does provoke some perspective-taking. "It's very intriguing to me that the universe has produced a very interesting period where a lot happens," she says. "And yet we seem to face a much longer period in utter darkness, cold. It's horrible. I feel, actually, from that point of view, very lucky, to be in the few years of cosmology where we are learning all this stuff for the first time."

"It momentarily makes me feel sad," agrees Andrew Pontzen. "Then I very quickly start worrying about our problems here on Earth right now and think, 'Come on.' We're in so much deeper trouble than the Heat Death of the universe. So I suppose it makes me start thinking about the problems that we face as a civilization on much shorter timescales. If I'm going to worry about anything, it's gonna be those, not the Heat Death."

"I just don't really have an emotional connection, I suppose, to the death of the universe," Pontzen continues, "but I do to the death of the Earth. I don't mind the fact that I'm going to die in fifty years or whatever, but I don't want the Earth to die in fifty years."

I have a lot of sympathy for this view. In terms of things we should actually worry about, the Heat Death, or vacuum decay, or the Big Rip, or whatever, cannot be at the top of that list (even setting aside the fact that we are utterly powerless to do anything about them). As living beings, we naturally care most about our own lives, and the lives of those close to us in space and time, and for the most part we leave the unimaginably distant cosmic future to its own devices.

But personally, I still feel there's a big difference, in some emotional sense, between "we go on forever" and "we don't." Nima Arkani-Hamed feels the same way. "At the absolute, absolute deepest level . . . whether or not people explicitly admit to thinking about it or not (and if they don't they're all the poorer for it) . . . If you think there is a purpose to life, then I at least don't know how to find one that doesn't connect to something that transcends our little mortality," he tells me. "I think a lot of people at some level—again, either explicitly or implicitly—will do science or art or something because of the sense that you do get to transcend something. You touch something eternal. That word, eternal: very important. It's very, very, very important."

Freeman Dyson had hoped to find a way to preserve intelligent life for all time. His 1979 paper proposed a way to propagate some kind of intelligent machine into an infinite future, through a scheme involving the constant slowing of processing and intermittent hibernation. Unfortunately, those calculations were done under the assumption that the universe's expansion doesn't accelerate, and now, it appears that it does. And if the acceleration continues, Dyson's plan won't work. "It would be disappointing," he admits. "I mean, you have to accept what nature provides. It's like the fact that we have finite lifetimes. It's not so tragic. In many, many ways it makes the universe more interesting. It's always evolving to something different. But having a finite lifetime for the whole thing . . . maybe that's our fate. But certainly, I would prefer to have evolution going on forever."

And who knows? Maybe there's a sense in which it does. Roger Penrose thinks there's a better way. He's spent the last decade or so developing his Conformal Cyclic Cosmology, which postulates a universe cycling from Big Bang to Heat Death, over and over again, forever, with the tantalizing possibility that something—some imprint from a previous cycle—might make it through the transition. The notion that what passes through could contain meaningful information about any conscious inhabitants is just idle speculation at the moment, he says, but the implications of that possibility could be profound. "I'm not certainly saying that I think this, but in some ways I find it less depressing . . . that maybe after one's death, it's conceivable there could be some legacy."

Or maybe the possibility of a multiverse landscape can comfort us. Jonathan Pritchard, a cosmologist at Imperial College London whose work has run the gamut from cosmic inflation to the evolution of galaxies, finds hope in the idea that in some other distant, unconnected region, something might exist long after we're nothing but waste heat. "Somewhere out there, there is a multiverse where stuff is always going on," he says. "Emotionally, I like the idea of that."

But *we* still die, I say.

He's unfazed. "It's not all about us, you know."

If we don't get to join the eternal multiverse party ourselves, at least our looming death can be good for physics. Neil Turok points out that the prospect of an end of time in the future, combined with the existence of our cosmic horizon, places hard boundaries on the universe, and thus helpful limits to the problem of understanding it all. A light wave traveling across a limited, expanding, accelerating universe can undergo only so many oscillations, even into the infinite future. "We live, effectively, in a box, okay? Which is finite. And if that's true, I think it's to be welcomed because we could understand it. The problem of understanding the universe just got a whole lot easier because it's finite," he says. "Finite to the past, finite in space because of the horizon, finite to the future because everything will only oscillate a finite number of times. Wow! I mean, that's understandable. I'm an optimist by nature, but I think the world is our oyster."

If the universe is going to end, one way or another, I concede that we may as well make our peace with it. Pedro Ferreira is way ahead of me on that one. "I think it's great," he says. "It's so simple and so clean.

"I've never understood why people get so depressed about the end, the death of the Sun and all," he continues. "I just like the serenity of it."

"So it doesn't bother you that we ultimately have no legacy in the universe?" I ask him.

"No, not at all," he says. "I very much like our blip-ness . . . It's always appealed to me," he continues. "It's the transience of these things. It's the doing. It's the process. It's the journey. Who cares where you get to, right?"

I admit it, I still care. I'm trying not to get hung up on it, on the ending, the last page, the end of this great experiment of existence. *It's the journey*, I repeat to myself. It's the journey.

There is perhaps some solace in the fact that whatever happens, it's not our fault. Renée Hložek considers this a definite plus.

"I love the fact that my work, even if I do it 100 percent perfectly and I'm an incredible scientist, it changes nothing about the fate of the universe," she says. "All we are trying to do is understand it. And even if you do understand it, we can do nothing to change it. I think that's freeing rather than scary."

To Hložek, the Heat Death isn't depressing, or boring. She calls it "cold and beautiful." "It's like the universe just sorts itself out," she says.

"What I hope that people get out of your book is that it's possible for the human mind to use observations of light—and/or gravitational waves, but let's stick with light for now—and make incredible inferences with relatively simple mathematics about the picture of the universe," Hložek says. "And even if we can do nothing to change it, that knowledge . . . even if that knowledge goes away, if all humans die, that knowledge right now is incredible. That's basically why I do what I do."

I think I see what she's saying. Would I want to uncover the secrets of the universe, even if I didn't get to share that knowledge, or keep it? I would. This seems important. "There's some purpose to doing it, even if it is lost."

"Because it changes who you are now, right?" she agrees. "I am delighted that we get to live at a time in the universe when we can see dark energy and not be ripped apart by it. But that means the whole point is that you understand it, and then you enjoy it, and then . . . 'so long and thanks for all the fish.' Cool."

Cool.

Acknowledgments

I never really imagined I would be an author, and I never would have managed it, if it were not for the help of far more people than I can name. I'm going to make an attempt to list just a small subset of those people here, but over the last few years I have taken much more support and advice than I could ever give back from countless friends and colleagues. If you are one of those people, whether your name appears here or not, please accept my thanks for all you have done and know that this book is partly yours as well. (I hope you like it!)

When I first set out to write this book, I had only a vague idea that I could put some words on paper and someone would, hopefully, eventually read them. Fortunately, I've been deftly guided through the whole process by my wonderfully patient, professional, and encouraging literary agent, Mollie Glick, and a whole team of enthusiastic book-crafters at Scribner. I'm especially grateful to Daniel Loedel for the feedback and edits that substantially honed and shaped this manuscript, and to Nan Graham for believing in my ability to write it in the first place. Thanks also to Sarah Goldberg, Rosaleen Mahorter, Abigail Novak, and Zoey Cole at Scribner, and Casiana Ionita, Etty Eastwood, and Dahmicca Wright at Penguin UK, who have all been working tirelessly for the last few months to get this book out into the world. I'm grateful to Nick James for the wonderful illustrations that appear in these pages and to Laurel Tilton and Ana Gabela for organizational support.

One of the biggest joys of this whole process has been having an excuse to connect and talk science with a huge number of amazing physicists and astronomers who have influenced the way I think about the cosmos. For indulging me and my many, many questions, I am grateful to Andy Albrecht, Nima Arkani-Hamed, Freya Blekman, Sean Carroll, André David, Freeman Dyson, Richard Easther, José Ramón Espinosa, Pedro Ferreira, Steven Gratton, Renée Hložek, Andrew Jaffe, Clifford V. Johnson, Hiranya Peiris, Sterl Phinney, Roger Penrose, Andrew Pontzen, Jonathan Pritchard, Meredith Rawls, Martin Rees, Blake Sherwin, Paul Steinhardt, Andrea Thamm, and Neil Turok. For volunteering to look over various chapters and give me extremely helpful feedback, I'm additionally indebted to several of the above and to Adam Becker, Latham Boyle, Sébastien Carrassou, Brand Fortner, Hannalore Gerling-Dunsmore, Sarah Kendrew, Tod Lauer, Weikang Lin, Robert McNees, Toby Opferkuch, and Raquel Riberio. Whatever errors are still in the manuscript (and I'm sure there are many) originate in my own failure to reliably commit the considerable collective wisdom of all of the above to the page.

While physicists may have borne the brunt of my technical queries, I have spent much of the last two years endlessly pestering nearly everyone I know with questions, drafts, advice requests, anxieties, and a general obsessiveness with all things book-related. I am deeply grateful to my friends and family for their patience, and to all the authors I know for lending me their perspectives on the writing and publishing world. Thank you to my family (especially my mom and my sister Jennifer) for encouraging and supporting me all my life and for letting me fill all our family gatherings with science and book talk. Thanks to Mary Robinette Kowal for writing tips and title ideas; Doron Weber for supporting my venture into this new space of public engagement; Daniel Abraham, Dean Burnett, Monica Byrne, Brian Cox, Helen Czerski, Cory Doctorow, Brian Fitzpatrick, Ty Franck, Lisa Grossman, Robin

Ince, Emily Lakdawalla, Zeeya Merali, Rosemary Mosco, Randall Munroe, Jennifer Ouellette, Sarah Parcak, Phil Plait, John Scalzi, Terry Virts, Anne Wheaton, and Wil Wheaton for extremely helpful book-writing advice; Charlotte Moore, Brian Malow, and the LA Nerd Brigade for endless encouragement and idea-bouncing; and Andrew Hozier Byrne for both inspiration and a killer soundtrack.

As a pre-tenure professor, I would not have dared to even begin this project were it not for the support of North Carolina State University, whose innovative Leadership in Public Science Cluster program made it possible for me to carve out an academic path that makes space for connecting with the public. The Physics Department and College of Sciences have been wonderfully supportive, helping me to find ways to balance the roles of author, researcher, mentor, and instructor.

Researching this book gave me the opportunity to travel to a number of institutions to interrogate my fellow physicists and to gain a new perspective on what this whole endeavor is all about. I'm particularly grateful to the people at CERN, the Institute for Advanced Study, the Perimeter Institute, the Aspen Center for Physics, Imperial College London, University College London, the Kavli Institute for Cosmology at Cambridge, and Oxford's Beecroft Institute for their hospitality during my visits.

And finally, special thanks to the wonderful staff of Jubala Coffee on Hillsborough Street, where the bulk of this manuscript was written. Your green tea and oatmeal gave me life.

Index

INDEX

INDEX

light
  dominant energy condition and, 111
  measuring shifts in. *See* blueshift measurements; redshift measurements
  movement through spacetime by, 20 (fig.)
  particle horizon and speed of, 82
  recession speed of galaxies and speed of, 83–84
  shape of space and response of, 68
  spectrum pattern characteristics of, 27, 55–57
  thermal radiation and, 27–28
  travel times for, 17 (fig.)
light speed delay
  description of, 16–17
  "now" concept when viewing events and, 18–19
  spacetime property and, 18–19
light-year unit, 16–17
LIGO. *See* Laser Interferometry Gravitational-Wave Observatory
Local Group of galaxies 52–53, 88
lookback time, 26
loop quantum gravity, 160
LSST. *See* Large Synoptic Survey Telescope

Many Worlds interpretation of quantum mechanics, 169n
matter
  bending of space by, 68, 108
  Big Bang Nucleosynthesis and, 45
  dark energy and, 107–8
  density over time of, 97 (fig.)
  imbalance between antimatter and, 162
  quark era and distinction between antimatter and, 44
maximum entropy universe, 97
McNees, Robert, 153n
Mendeleev, Dmitri, 137

Mercury
  gravity theory and observations of, 8, 162
  red giant phase of the Sun and destruction of, 1, 121
Milky Way, 8, 24
  Andromeda Galaxy's future collision with, 51–52, 62–63, 88
  distance ladder measurements and, 119 (fig.)
  smaller nearby galaxies consumed by, 52
"Mixed Signals" (White and Wharton), 166
Moon
  Big Rip and, 113
  cosmic ray collisions on, 132–33
  distance and apparent size of, 86
Moss, Ian, 150

neutron, decay of, 89
neutron stars, 125n, 158, 159, 166, 171
Newton, Sir Isaac, 8, 9
Nietzsche, Friedrich, 3, 100–101

observable universe
  description of, 21, 26
  cartoon map of, 26 (fig.)
  cosmic expansion and, 54
  particle horizon in, 82–84
  seeing "edge" of, 83
  uniformity problem of, 39–40
open universe, 75, 76 (fig.)
overview effect, 7

parallax, in distance measurement, 117, 118, 119 (fig.)
Parkes radio telescope, 24
particle colliders, 48. *See also specific colliders*
  early universe research using, 134–35
  gravity theory experiments and, 161

222

About the Author

Katie Mack is a theoretical astrophysicist exploring a range of questions in cosmology, the study of the universe from beginning to end. She is currently an assistant professor of physics at North Carolina State University, where she is also a member of the Leadership in Public Science Cluster. Alongside her academic research, she is an active science communicator and has been published in *Scientific American*, *Slate*, *Sky & Telescope*, *Time*, and *Cosmos Magazine*, where she is a columnist. She can be found on Twitter as @AstroKatie.